Statistical Modelling and Analysis of Fused Deposition Modelling Process Parameters for Green Composites Fabrication

This book offers a comprehensive exploration of the intricacies involved in conducting statistical analyses on Fused Deposition Modelling (FDM) printed green composites, backed by real-world examples. It helps readers navigate the statistical terrain within the realm of eco-friendly composite materials, offering a hands-on approach that bridges theory with tangible applications.

Key features:

- Contains processing and manufacturing of novel green filaments for additive manufacturing (AM) processes
- Includes experimental, analytical, and numerical analyses of novel green composite filaments for the 3D printing industries
- Covers statistical analysis for green composites development using AM
- Deals with mechanical, thermal, and machinability of green composite filaments
- Addresses green filaments in various AM applications and their challenges

This book is aimed at researchers and graduate students in composites, polymers, modelling, manufacturing, and materials engineering.

Statistical Modelling and Analysis of Fused Deposition Modelling Process Parameters for Green Composites Fabrication

Sathish Kumar Adapa,
Jagadish and Divya Zindani

CRC Press
Taylor & Francis Group
Boca Raton London New York

CRC Press is an imprint of the
Taylor & Francis Group, an **informa** business

First edition published 2026
by CRC Press
2385 NW Executive Center Drive, Suite 320, Boca Raton FL 33431

and by CRC Press
4 Park Square, Milton Park, Abingdon, Oxon, OX14 4RN

CRC Press is an imprint of Taylor & Francis Group, LLC

© 2026 Sathish Kumar Adapa, Jagadish and Divya Zindani

ISBN: 9781032796574 (hbk)
ISBN: 9781041224112 (pbk)
ISBN: 9781003732143 (ebk)

DOI: 10.1201/9781003732143

Typeset in Times
by codeMantra

Contents

About the Authors

Sathish Kumar Adapa is working as an Assistant Professor in the Department of Mechanical Engineering, AITAM College, Andhra Pradesh, India. He has completed his M. Tech. in Design and Manufacturing from CUTM, Odisha, India, and is currently pursuing a PhD at NIT Raipur, India. During his nine years of experience, he has published more than ten international and national journal articles, conferences, and book chapters. His research interests include additive manufacturing, green composites, green manufacturing, advanced manufacturing process, CAD/CAM, and composite material and machining characterization.

Jagadish is working as an Assistant Professor in the SQC & OR Unit, Indian Statistical Institute, Bangalore, India. Prior to this, he worked as an Assistant Professor in the Department of Mechanical Engineering at NIT Silchar & NIT Raipur. He completed his PhD in Industrial Engineering from NIT Silchar, India. During his 12 years of experience, he has published more than 98 articles in international and national journals, conferences, and book chapters. He has received an Institutional Award (Gold Medal) by the Institution of Engineers India, Best Paper Award by NIT Silchar, and Best Research Paper Award by NIT Raipur for his outstanding research contribution. His research interest includes industrial engineering, optimization techniques, supply chain management, time series analysis, operations management, statistical modelling and analysis, six sigma, quality management, design of experiments; statistical quality control, green composites, and green manufacturing.

Divya Zindani received his PhD degree in Mechanical Engineering from the National Institute of Technology Silchar, Assam, India, in 2021. He is currently working as an Assistant Professor in the Department of Mechanical Engineering, Sri Sivasubramaniya Nadar College of Engineering, Chennai. He has over two years of industrial experience. His areas of research interest are decision-support systems, artificial intelligence/machine learning, optimization, sustainable product development and process design, and prognostic health monitoring. He has more than 15 patents, more than 15 authored and edited books, 20 book chapters, and 30 international journal publications to his credit. He has been felicitated with many awards. He is an active member of Multiple Criteria Decision Making, The International Society.

Preface

The increasing global urgency to address environmental degradation, resource depletion, and climate change has placed sustainability at the forefront of materials science and manufacturing innovation. As industries and researchers strive to adopt eco-conscious alternatives, green composites, biodegradable polymers reinforced with natural fibres, have emerged as viable, sustainable materials capable of transforming product development across domains. However, their full potential can only be realized through a systematic and scientific understanding of their processing behaviour and mechanical performance.

This book, *Statistical Modelling and Analysis of Fused Deposition Modelling Process Parameters for Green Composites Fabrication*, is a focused effort to bridge material science, AM, and statistical engineering. It serves as a comprehensive resource that brings together theoretical underpinnings, experimental investigations, and advanced modelling techniques to support sustainable composite development.

This book is structured into six chapters. Chapter 1 introduces the reader to green composites and their application domains, providing historical context and highlighting the urgent need for sustainable alternatives. Chapter 2 offers a deep dive into statistical modelling tools: design of experiments (DOE), response surface methodology (RSM), analysis of variance (ANOVA), regression techniques, principal component analysis (PCA), and decision-support systems, which are essential for optimizing and predicting the behaviour of FDM-processed green composites.

Chapters 3–6 explore material-specific investigations involving jute, coir, bamboo, and cane fibres, reinforced into polylactic acid (PLA) matrices and fabricated using FDM. Each chapter details the design, processing, mechanical characterization, and statistical optimization of these composites. Experimental strategies such as Taguchi methods and L_9 orthogonal arrays are used to derive meaningful insights into parameter tuning and performance enhancement.

This book is intended for researchers, graduate students, and practitioners in mechanical engineering, materials science, and AM who are invested in developing sustainable technologies. The interdisciplinary nature of the text also makes it valuable for academicians and industry professionals seeking data-driven methodologies for bio-based materials.

We hope that the methodologies and insights offered here inspire further research, innovation, and adoption of sustainable composite technologies in real-world applications.

Sathish Kumar Adapa
Dr. Jagadish
Dr. Divya Zindani

Acknowledgements

With deep gratitude and humility, we wish to express our sincere thanks to all those who have supported us in the successful completion of this book, *Statistical Modelling and Analysis of Fused Deposition Modelling Process Parameters for Green Composites Fabrication.*

First and foremost, we bow in reverence to the Almighty, whose divine grace, strength, and wisdom have guided us throughout this journey. Without His blessings, this work would not have been possible.

We express our heartfelt thanks to our parents and families, whose unwavering encouragement, patience, and moral support have been our constant source of inspiration.

We extend our sincere appreciation to the publisher for believing in the merit of this work and facilitating its dissemination to a wider academic and industrial audience.

We gratefully acknowledge the support of Aditya Institute of Technology and Management, Tekkali, Srikakulam District, Andhra Pradesh, for generously providing access to the Thermal Engineering Laboratory facilities, which were instrumental in the fabrication of our composite specimens. Their infrastructure and technical assistance formed the backbone of our experimental investigations. We also extend our deepest gratitude to GITAM, Visakhapatnam, Andhra Pradesh, for their timely support and cooperation in testing and evaluating the composite samples. Their invaluable contributions have played a crucial role in the validation and success of this work.

To all mentors, colleagues, friends, and institutions who directly or indirectly contributed to this endeavour, we offer our sincere thanks.

We dedicate this work to the pursuit of sustainable innovation and knowledge for the betterment of society.

Abbreviations

ABS	acrylonitrile butadiene styrene
AC	*Acacia concinna*
AHP	Analytic Hierarchy Process
AI	artificial intelligence
AM	additive manufacturing
ANNs	artificial neural networks
ANOVA	analysis of variance
ARD	absolute relative deviation
ASTM	American Society for Testing and Materials
BBD	Box–Behnken design
BT	bed temperature
CAD	computer aided design
CCD	central composite design
CF	carbon fibre
CO$_2$	carbon dioxide
DF	degrees of freedom
DIY	do-it-yourself
DOE	design of experiments
DSS	decision-support systems
E	Young's modulus
FDM	fused deposition modelling
FL	fibre loading
GA	genetic algorithm
GRA	Grey Relational Analysis
ID	infill density
LCA	life cycle analysis
LT	layer thickness
MCDM	multi-criteria decision-making
ML	machine learning
MLR	multiple linear regression
MS	mean square
NaOH	sodium hydroxide
NT	nozzle temperature
OAs	orthogonal arrays

PBAT	polybutylene adipate terephthalate
PBD	Plackett–Burman design
PBS	polybutylene succinate
PC	principal component
PCA	principal component analysis
PCL	polycaprolactone
PETG	polyethylene terephthalate glycol
PHA	polyhydroxyalkanoate
PHB	polyhydroxybutyrate
PLA	polylactic acid
PNCs	Particle Number Concentrations
PP	polypropylene
PSO	particle swarm optimization
RMSE	root mean square error
RP	rapid prototyping
RSM	response surface methodology
RTM	resin transfer moulding
S/N Ratio	signal-to-noise ratio
SA	simulated annealing
SCFs	small coir fibres
SEM	scanning electron microscopy
SERs	Specific Emission Rates
SSA	Sum of Squares for Factors
SSAB	Sum of Squares for Interaction
SSE	Sum of Squares for Error
SST	Total Sum of Squares
SVMs	support vector machines
TGA	thermogravimetric analysis
TOPSIS	Technique for Order Preference by Similarity to Ideal Solution
TPs	thermoplastics
UFPs	ultrafine particles
UTS	ultimate tensile strength
UV light	ultra violet Light
VIKOR	VlseKriterijumska Optimizacija I Kompromisno Resenje
VOCs	volatile organic compounds

Introduction

The global shift towards sustainability has created a pressing need for innovative materials and manufacturing technologies that reduce environmental impact while maintaining high performance and functional adaptability. Among these, green composites, comprising biodegradable polymers reinforced with natural fibres, have emerged as a transformative class of materials for next-generation, eco-friendly engineering applications. Their potential spans across automotive, aerospace, biomedical, consumer products, and packaging industries, aligning with the broader goals of the circular economy and green innovation.

Simultaneously, additive manufacturing (AM), particularly fused deposition modelling (FDM), has gained prominence as a versatile and resource-efficient technique for fabricating complex geometries with minimal material wastage. When combined, green composites and FDM offer an exciting avenue for sustainable product development, enabling personalized, low-footprint manufacturing using renewable and biodegradable resources. However, the use of natural fibres and bio-based polymers introduces challenges such as variability in material properties, non-uniform process behaviour, and a high degree of parameter sensitivity.

To address these challenges and optimize the fabrication of green composites via FDM, this book presents a comprehensive exploration of statistical modelling and data-driven analysis techniques. By integrating methods such as design of experiments (DOE), response surface methodology (RSM), ANOVA, regression analysis, Taguchi methods, and principal component analysis (PCA), this book offers systematic tools to decode the complex relationships between processing parameters and material performance.

Spanning six detailed chapters, this book begins with foundational insights into green composite materials, highlighting the types of natural fibres and biodegradable matrices commonly used. Subsequent chapters delve into statistical modelling frameworks that support process optimization and predictive modelling. Case studies centred on jute, coir, bamboo, and cane fibre-based PLA composites showcase the application of these methods in real experimental settings. Each case study emphasizes the formulation of empirical models, optimization of mechanical properties, and validation through confirmatory experiments.

What sets this book apart is its focus on the integration of statistical models within sustainable design and decision-support systems, allowing engineers and researchers to align performance metrics with environmental considerations. This holistic, data-centric approach to materials engineering is essential for driving innovation in a resource-constrained world.

This book is intended for:

- Graduate students and researchers in materials science, mechanical engineering, and polymer engineering.
- Industry professionals involved in sustainable product development and AM.
- Academics and practitioners seeking practical tools for experimental design, analysis, and optimization of bio-composite materials.

By bringing together theory, application, and sustainability, the authors hope this book serves as a valuable reference for building the next generation of green, optimized, and intelligent material systems.

Introduction to Green Composites and Their Applications

1

1.1 INTRODUCTION

Green composites or bio-composites represent a new group of materials that seek to lessen the ecological impact conventionally caused by composite materials [1–3]. These consist of natural fibres like jute, flax, hemp, kenaf, sisal, bamboo, and coir that are reinforced using biodegradable or bio-based resins. The resins may either be made up of renewable sources such as vegetable oils, starch, and other natural plant-based polymers or formulated to be biodegradable in nature, which will automatically disintegrate with the passage of time. The most important concept of green composites is to utilize sustainable and renewable resources without relying excessively on synthetic, petroleum-derived materials, causing harm to the environment.

The increasing demand for green composites is due to the escalating need for environmentally friendly solutions in material science across the globe. The world is threatened by extreme environmental issues such as resource depletion, pollution, and climate change. Green composites provide a way to shift to greener alternatives. They advocate for the use of materials that leave the least possible ecological footprint, from raw material extraction to

DOI: 10.1201/9781003732143-1

disposal. Their use of naturally occurring fibres and biodegradable polymers makes them the best choice where sustainability is of major concern, e.g., in automotive components, packaging, building, and consumer products. By drawing on renewable materials, green composites can assist in the alleviation of pressure on non-renewable resources, minimize greenhouse gas emissions, and decrease end-of-life waste [4].

The green composites have extensive environmental advantages that make them very desirable as alternatives to traditional composites. Reduced carbon footprint related to the usage and manufacturing of these materials is one of the most significant advantages [5]. Natural fibres, for instance, are products of plants that fix carbon dioxide (CO_2) while they grow. That implies that green composites lead to carbon sequestration and compensating emissions resulting from the manufacture and processing of the composite materials. Green composites, if their whole life cycle, ranging from the stage of growing raw materials through decay, produces fewer net releases of CO_2 as compared to their conventional composite equivalents, usually the synthetic counterpart used in them derives from fossil sources.

In addition, green composites are usually biodegradable, implying that they may degrade naturally in the presence of environmental factors such as moisture and microorganisms. This aspect reduces the volume of waste sent to landfills tremendously. Conventional composites, however, are produced using non-biodegradable materials such as fibreglass and carbon fibre reinforced by petroleum-based resins [6,7]. These materials are not easily degradable and can remain in the environment for hundreds of years, which adds to the world's waste problem. Green composites, on the other hand, can degrade in a relatively short period of time, depending on the material, lessening the environmental impact of disposal.

Another significant environmental advantage is that green composites are sustainable on the premise that they are produced from renewable resources. Fibres from natural sources are typically locally obtained, cheap, and require less energy compared to synthetic fibres. Plant crops such as jute, hemp, and flax typically will consume less energy than synthetic fibres, which are high-energy chemical products. Also, most of the natural fibres employed in green composites are crop side products, i.e., they are making effective use of waste. It fosters circular economy because it's recycling material continuously, hence lesser demand for virgin material.

Green composite production and processing are designed to consume less energy compared to conventional composites. Green composites are manufactured at lower energies and with fewer chemicals, reducing the energy and environmental expenses of production. Traditional composites, however, require high temperature and energy-intensive processes such as polymerization and curing to attain their end properties. Green composites' low energy needs translate to reduced greenhouse gas emissions during production, a factor that enhances overall sustainability.

Compared to conventional composites, several of their commonalities with conventional composites are identified, particularly environmental impact and performance properties. Conventional composites, conventionally produced from raw materials such as glass fibres, carbon fibres, and aramid fibres, are conventionally reinforced with man-made resins such as epoxy, polyester, or vinyl ester. Raw materials are petrochemical-based and hence non-renewable and non-biodegradable. The manufacturing process of conventional composites is energy-intensive and liberates large quantities of CO_2 and other greenhouse gases. The conventional materials are also resistant to recycling and therefore pose long-term environmental challenges in their elimination.

Green composites, nonetheless, have superior environmental performance from the application of natural fibres as well as biodegradable resins. Since green composite fibres are natural resources, they have a little environmental impact on their entire lifespan. Natural fibres, for example, take less energy to manufacture and produce than synthetic fibres, and can be produced using sustainable farming [8,9]. In addition, the green aspect of biodegradable green composites is that they won't add to the exploding plastic waste problem, one of the biggest failings of conventional composites.

For performance, though, there are some compromises to be made between green composites and conventional composites. While green composites should be lighter and cleaner, they are not necessarily equivalent to conventional composites in mechanical performance. For instance, carbon fibre composites exhibit higher tensile strength, stiffness, and thermal resistance, and thus are more appropriate for aerospace and automotive applications where such high-strength material is needed. Natural fibre composites, however, can be weaker and stiffer than their counterparts, but research is currently directed towards enhancing the latter by way of better fibre treatments, hybridization, and new bio-based resin development.

The strength and moisture resistance of green composites are other key aspects that need to be tested. Natural fibres are hydrophilic in character; i.e., they readily absorb water compared to synthetic fibres, which is a disadvantage to the mechanical properties of the composite in the long term. This disadvantage renders green composites less suitable to be used in sea or high-humidity environments unless treated specially or compounded with resins for moisture resistance. They are more environment-resistant traditionally, especially glass or carbon fibre composites, therefore inherently a more logical option for a long period at severe conditions.

Yet, with advancements in material science, the gap between green and conventional composites is gradually bridging. Researchers are looking towards hybrid composites, where synthetic strengthening agents are added to natural fibres to benefit from the virtues of both worlds: environmental responsibility and enhanced mechanical strength [10,11]. New advancements in bio-based resins and surface treatments on the fibre are also making green composites moisture-resistant and stress-resistant in nature, opening doors to mass applications.

1.2 HISTORICAL BACKGROUND

The idea of green composites cropped up because of increased environmental consciousness and the demand for eco-friendly options to replace man-made materials towards the latter part of the 20th century. While the construction use of natural fibres and other uses have been in existence for thousands of years (for example, constructing houses from straw and mud), the concept of using natural fibres mixed with resins to create composite materials only started in the 1960s and 1970s.

One of the earliest achievements in green composite development was the greater use of natural fibres such as flax, jute, hemp, and sisal as reinforcing materials. All of these had been prized for their strength, flexibility, and availability for a long time, but lacked the technology to effectively pair them with resins. The development of the environmental movement in the 1960s also played an important role in shaping interest in sustainable materials, prompting scientists to find alternatives for the conventional glass fibre and carbon fibre composites using natural fibre-based composites.

In the early 1990s, there was a significant advancement with the creation of biodegradable polymers from renewable feedstocks such as corn, potato starch, and sugarcane. These bio-based resins provided an alternative means to manufacture composites that not only had natural fibres as reinforcement but were also entirely biodegradable.

As environmental concerns gained momentum across the decades, green composites continued to grow and evolve, both in terms of material and manufacturing process. One of the principal areas of research was in the creation of bio-based resins. The early resins, although biodegradable, tended to lack performance properties such as strength and durability. Through the passage of time, advancements in polymer chemistry brought about increased-performance, more functional bio-based polymers such as polylactic acid (PLA), polyhydroxyalkanoates (PHAs), and bio-polyesters. These bio-resins gave green composites higher mechanical properties and were competitive with conventional composites to compete on performance-critical applications.

In parallel, processing of natural fibres also improved. Natural fibre composites in the early years were plagued by issues of fibre-matrix bonding and moisture absorption, which constrained their strength and durability. To mitigate these problems, scientists started to create surface treatments for natural fibres, including chemical treatments, to improve the bonding with the matrix material and the resistance to moisture. The hybrid use of materials—mixing synthetic fibres or resins with natural fibres—also became widely used, enabling enhanced mechanical properties while some level of sustainability was preserved.

Another area where significant development has occurred is in the fabrication process. Standard processes for composites fabrication, like hand lay-up, resin transfer moulding (RTM), and compression moulding [12,13] have been tailored according to the unique needs of natural fibre-reinforced composites. New methods were evolved over a period to enhance the efficiency, consistency, and quality of green composites production. For instance, procedures like vacuum infusion and automated fibre placement were developed to be compatible with bio-materials to produce more intricate and high-performance green composites on large scales.

The automotive and packaging sectors [14] have been key driving forces in the development of green composites. In the early 2000s, auto manufacturers such as Ford and Toyota started investigating the use of green composites for interior and body panel applications in cars as part of overall efforts to make cars lighter and more sustainable. Ford, for example, generated headlines by using soy-based foam and natural fibre-reinforced plastics in some vehicle models. These initial industrial applications catalyzed more research on how to enhance the performance and scalability of green composites.

Green composites are evolving today with constant research to enhance their mechanical properties, biodegradability, and affordability. Nanotechnology and the use of nano-fillers are creating new avenues for green composites by making them stronger, thermally stable, and resistant to environmental conditions. The establishment of circular economy models is also promoting recycling and reuse of bio-based materials, further boosting the sustainability value of green composites.

1.3 MATERIALS USED IN GREEN COMPOSITES

Green composites, also known as sustainable composites, are largely based on natural fibres and biodegradable resins, which classify them as environmentally friendly composites as compared to the traditional synthetic composites. They have the most significant importance in lowering the ecological footprint of industries like the automotive, building, packaging, and consumer industries.

1.3.1 Natural Fibres in Green Composites

Among the natural reinforcement materials in green composites, jute, cane, bamboo, and coir fibres are notable because they are widely available, renewable, biodegradable, and possess good mechanical properties. Natural fibres are becoming widely known for the potential to replace synthetic fibres such as glass and carbon for use in composite materials. Natural fibres are light in weight, can be renewed, and can be harvested sustainably. Following is a comprehensive description of four major natural fibres, jute, cane, bamboo, and coir, and their characteristics and benefits.

1.3.1.1 Jute Fibres

Jute is extremely durable, versatile, and easy to cultivate among the most used natural fibres in the world. Jute fibres are derived from the plant species *Corchorus capsularis* and *Corchorus olitorius*, cultivated primarily in the tropics like India, Bangladesh, and China. Jute fibres possess average tensile strength, with a common range of 200–800 MPa, depending on fibre quality and processing conditions [15]. The jute density is approximately 1.3 g/cm³, which is relatively low compared to glass fibres (2.5 g/cm³). Jute fibres have relatively high stiffness, with a Young's modulus between 10 and 30 GPa. A disadvantage of jute is that it has a high water absorption rate, which is usually around 12%–13%. This can impact the mechanical properties of the composite material over time. Jute is extremely renewable with a growth cycle of just four to six months. It is one of the most environmentally friendly fibres because it has low water and pesticide demands. Jute fibres are completely biodegradable, which makes them a perfect option for green composites. Jute is not very expensive compared to man-made fibres, which has led to its extensive application in composites for packaging, automotive parts, and non-structural building products.

1.3.1.2 Cane Fibres

Fibres of cane, particularly from sugarcane bagasse, are sugar industry by-products. Bagasse is fibrous matter after crushing stalks of sugarcane to get juice. Cane fibres are gaining attention over the past few years as an eco-friendly alternative for composite material. Tensile strength of cane fibres tends to range between 100 and 300 MPa, subject to processing and quality of the fibre. The density of cane fibres is around 1.25 g/cm³, which is lower compared to the majority of synthetic fibres [16]. Stiffness of cane fibres is relatively moderate with a Young's modulus of around 5–15 GPa. Cane fibres are prone to the absorption of moisture, and this may affect their mechanical characteristics in outdoor application. Since bagasse is a waste product of the sugarcane industry, composite utilization adds value to waste material and promotes circular economy. Cane fibres are fully biodegradable and ideal for short-term products like disposable packaging. With the global spread of sugar production, cane fibres are cheap and readily available.

1.3.1.3 Bamboo Fibres

Bamboo is a very fast-growing plant in the world, so it is a very renewable source of natural fibres. The fibres of bamboo are derived from bamboo species, and they are very popular as they possess excellent strength-to-weight ratio and also, they are beneficial for the environment. Bamboo fibres contain high tensile strength that is generally 400–800 MPa, so they are as strong as some of the synthetic fibres such as glass. Bamboo is relatively light in weight, with low density values ranging from 0.9 to 1.2 g/cm³ [17]. Bamboo fibres are stiffer than other conventional synthetic fibres, with values ranging from 10 to 40 GPa based on the species and treatment of the fibres. Bamboo also has greater water resistance than many natural fibres, such as jute and cane, but it must be appropriately treated to prevent water uptake. The unusually high tensile strength of bamboo, coupled with its lightness, makes bamboo appropriate for composites' load-bearing applications. Bamboo is remarkably fast-growing and, in certain species, as much as 3 ft/day. This also ranks it high on the scale of renewable natural resources. Being biodegradable, like any other natural fibre, bamboo reinforces its sustainability report.

1.3.1.4 Coir Fibres

Coir is a natural fibre obtained from coconut husk. It is durable, resistant to water, and bears harsh weather conditions. Coir fibres are found abundantly in tropical countries, especially in India, Sri Lanka, and the Philippines. Coir fibres possess a tensile strength of approximately 120–200 MPa, which is less than other natural fibres such as jute and bamboo but is adequate for most uses [18]. The coir fibres have a density of about 1.15 g/cm³, so they are light. Coir fibres are softer than other natural fibres, with a Young's modulus of 4–6 GPa. Coir is also highly water- and salt-resistant, so it is suitable for marine and outdoor applications.

Coir's inherent resistance to salt and moisture makes it a good option for use in marine conditions or outdoor construction. Coir is resistant to environmental stress and thus durable in many applications. Coir is a renewable resource and completely biodegradable, which adds to its application in sustainable composite materials.

1.3.2 Biodegradable Resins in Green Composites

Natural fibres are the reinforcement used in green composites, while biodegradable resins are the matrix that holds the fibres together and enables load transfer among the fibres. Biodegradable resins are polymer materials that are capable of degradation by the action of microorganisms such as bacteria and fungi as well as by natural environmental conditions like moisture, temperature, and sunlight. In contrast to conventional synthetic resins, which may last in the environment for centuries, biodegradable resins break down into more basic molecules, including water, CO_2, methane, and biomass, within a relatively short period of time. Biodegradable resins can be divided into two broad categories: bio-based biodegradable resins, which are made from renewable resources such as plant starches, vegetable oils, and natural sugars. Petroleum-derived biodegradable resins, which, while being derived from fossil fuels, have been particularly chemically designed to break down in the natural world. Utilization of biodegradable resins in green composites is in line with the world's sustainability objectives as it responds to plastic pollution issues and decreases reliance on non-renewable materials. The resins are typically applied together with natural fibres such as jute, flax, coir, cane, and bamboo to produce green composites that can be used in a wide range of applications, ranging from automotive parts to packaging materials. Some of the biodegradable resins have come up as great green composite materials owing to their exceptional properties and environmental compliance. PLA, PHA, polycaprolactone (PCL), polybutylene succinate (PBS), and starch-based polymers are the most used biodegradable resins. All these resins possess different strengths and are ideal for various composite uses.

1.3.2.1 Polylactic Acid

PLA is among the most common biodegradable resins used in green composites. PLA is obtained from renewable sources like corn starch, sugarcane, and cassava and has become increasingly popular because of its availability, biodegradability, and low cost of production. PLA is obtained by the polymerization of lactic acid, which is sourced via the fermentation of plant sugar. PLA degrades into water, CO_2, and organic matter under composting conditions. Under industrial composting conditions, PLA can degrade within months. PLA has good stiffness and strength and is therefore appropriate for structural uses. Its relatively low impact resistance and brittleness can be drawbacks in certain applications. PLA is readily processed with standard manufacturing methods like extrusion, injection moulding, and 3D printing, and it is thus versatile for a range of composite fabrication processes. PLA is one of the favoured candidates for green composites because it is renewable in nature and biodegradable. The combination of natural fibres with PLA leads to superior mechanical properties and lower environmental footprints. PLA-based composites are widely applied in packaging, automotive interior, and biodegradable consumer products.

1.3.2.2 Polyhydroxyalkanoates

PHAs are a group of naturally occurring biodegradable polyesters synthesized by a wide range of microorganisms during the process of fermentation. PHAs are synthesized using renewable carbon sources like plant oils and sugars, thus resulting in a completely bio-based and biodegradable material. The most popular forms of PHAs are polyhydroxybutyrate and polyhydroxyvalerate. PHAs are completely biodegradable in several environments, such as soil, water, and composting environments. They can degrade between weeks to several months, depending on environmental conditions. They have good mechanical properties, including high tensile strength and flexibility, and thus they are appropriate in applications where toughness is necessary. They are, however, typically more brittle compared to some of the other biodegradable resins. PHAs have a relatively low melting point, which restricts their application in high-temperature uses. PHA-based resins are also quite ideal for green composites due to their biocompatibility and hence can find application in medicine and packaging. PHA-based composites with natural fibres have increased biodegradability along with improved mechanical properties. Such composites have applications in making biodegradable packaging, farm films, and medical implants.

1.3.2.3 Polycaprolactone

PCL is a petroleum-derived biodegradable polymer with characteristics of flexibility, low melting point, and gradual degradation. Even though PCL is petroleum-derived, it can be broken down by the action of microorganisms and is eco-friendly in the sense that it degrades over time. PCL biodegrades slowly under normal conditions and will take many years to completely disintegrate. It will degrade more rapidly, though, under industrial composting conditions. PCL is extremely flexible with high elongation, making it good for uses needing ductility. It has, however, poorer tensile strength than some of the other biodegradable materials, such as PLA and PHA. The low melting point (approximately 60°C) of PCL makes it inappropriate for high-temperature applications. PCL is generally mixed with other biodegradable resins or filled with natural fibres to enhance its mechanical characteristics and extend its range of application. When used with fibres such as jute, coir, or bamboo, PCL-based composites find applications in medical devices, drug delivery, and biodegradable packaging.

1.3.2.4 Polybutylene Succinate

PBS is a biodegradable polyester produced through the polymerization of succinic acid and 1,4-butanediol, which may be produced from renewable materials. PBS has attracted interest because of its excellent thermal stability, biodegradability, and compatibility with most natural fibres. It breaks down into water and CO_2 in composting conditions in a matter of months and is thus an ideal material for green composites to be employed in short-life cycle applications. PBS has good toughness, impact resistance, and flexibility, making it an ideal material to be employed in applications calling for durability. Additionally, it possesses a relatively high melting point (about 115°C), which makes it applicable under a broader spectrum of temperature conditions than certain other biodegradable resins. PBS is also often blended with natural fibres such as jute, flax, and bamboo to form green composites that have improved mechanical properties and biodegradability. These green composites find widespread applications in packaging, agricultural films, and automotive interior parts.

1.3.2.5 Starch-Based Polymers

Starch polymers are biodegradable resins obtained from renewable starch sources like corn, potato, and tapioca. Starch is the most widely available renewable resource and is used widely in the manufacture of biodegradable plastics. Starch polymers are completely biodegradable, degrading into water, CO_2, and organic matter in a matter of months under composting conditions. Pure starch-based polymers are brittle and have poor mechanical strength. Still, if they are combined with other biodegradable polymers, including PLA or PCL, then their mechanical strengths could be enormously enhanced. Starch is an accessible and cost-effective material; thus, starch-based polymers represent an economically favourable approach towards the utilization of green composites. Starch-based polymers can be blended with natural fibres to produce biodegradable composites that are viable for low-cost applications, such as packaging products, disposable tableware, and agricultural items. The application of starch-based polymers in green composites lessens the dependency on petroleum plastics and provides an eco-friendly means for single-use products.

1.4 FABRICATION TECHNIQUES FOR GREEN COMPOSITES

The increasing need for green materials has created the possibility of green composites composed of natural fibres and biodegradable resins. Green composites provide a more environmentally friendly option to conventional synthetic composites without any loss of mechanical properties. As green composites are developed, manufacturing processes also evolve to make their production possible. Fabrication processes are instrumental in determining the performance, cost-effectiveness, and scalability of the material. Of these methods, fused deposition modelling (FDM) has emerged as a pioneering and adaptive process to create green composites, with more conventional ones such as compression moulding and injection moulding. These manufacturing processes are elaborately explained in this section, with specific focus on FDM and short discussions on other major processes.

1.4.1 Introduction to FDM

FDM is one of the most widely used additive manufacturing techniques and is especially suited for the manufacturing of green composites [19]. FDM was developed in the late 1980s and has since become a standard process for 3D printing. This is because it is user-friendly, cheap, and capable of printing intricate geometries. Though FDM is normally associated with thermoplastics, recent advancement has made it possible to apply it to bio-degradable polymers and natural fibre-reinforced filaments, and as such, it is the optimum way to produce green composites. In FDM, a thermoplastic polymer is heated to a melt and extruded out of a hot nozzle and layer by layer onto a build platform. The material solidifies to the required shape.

It can be modified to use biodegradable polymers such as PLA, PHAs, or bio-based polymers blended with natural fibres such as jute, flax, and bamboo to form green composites.

1.4.1.1 Process of FDM for Green Composites

The FDM of green composites proceeds on the same general basis as normal FDM but with special focus placed on biodegradable and natural fibre-reinforced filaments. The most important steps are:

- Material Choice: The initial step is to choose the right biodegradable materials, i.e., PLA or PHA, blended with natural fibres such as flax, jute, bamboo, or wood. These materials come in the form of filaments, which are supplied to the FDM machine.
- Computer-Aided Design (CAD) Model Preparation: A 3D CAD model of the object to be printed is created with the help of CAD software. The model is then translated into a format that can be read by the FDM machine (typically an STL file).
- Toolpath Generation and Slicing: The 3D model is "sliced" into very thin layers with the help of dedicated slicing software. The software creates a toolpath for the FDM machine, telling it where to deposit the material in a layer-by-layer fashion.
- Biodegradable Material Extrusion: The biodegradable composite filament is inserted into the FDM printer hot extruder. The filament is melted by the hot nozzle, and according to the toolpath, the material is extruded.
- Layer-by-Layer Deposition: The extruder travels along the X, Y, and Z axes to deposit material layer by layer. A layer is cooled and solidified first before the next one is deposited on top of it, building up the object layer by layer.
- Cooling and Solidification: After the last layer has been deposited, the material solidifies and cools to create the final product. Depending on the finish smoothness that is required, some of the finishing processes, such as polishing or sanding, may be required.

FDM has several important benefits when utilized to produce green composites, such as:

- Complex Geometries and Tailoring: Complex geometries are produced using FDM that are difficult with traditional techniques. This is useful when the design has complex structures.

- Material Efficiency: FDM is an additive process; i.e., material is deposited where it is required. This minimizes loss of material, and hence the process is cost-effective and environmentally friendly, especially with the use of costly bio-based materials.
- Flexibility: FDM can accommodate a broad group of biodegradable polymers as well as natural fibre-strengthened filaments with ease, yielding flexibility in the choice of the material. This allows one to produce green composites of stated mechanical performance according to their intended application.
- Low-Cost Manufacturing: FDM technology is more affordable than other additive manufacturing technologies. FDM is thus affordable to a wide range of industries, including those with limited budgets or small production quantities.
- Rapid Prototyping: FDM finds its best application in rapid prototyping, where designers and engineers can create and test composite components within a brief period and later produce them in large numbers. This is of great use in research and development centres where perhaps numerous iterations would be required.
- Sustainability: FDM minimizes the environmental footprint using biodegradable materials and natural fibres. Recycling of the prints that have failed or the excess filament is another area that enhances its sustainability. FDM is being increasingly adopted in industries that are looking to incorporate sustainability into their manufacturing processes. Some key application areas for FDM-printed green composites include:
- Automotive: FDM-manufactured green composites find use in the production of interior parts, panels, and electric and hybrid automobile light parts. It is particularly worth it here because it has the capability of quickly and greenly manufacturing intricately customized components.
- Consumer Goods: FDM is used to fabricate biodegradable and natural fibre-reinforced products such as household items, packaging, and furniture components. These products offer an eco-friendly alternative to conventional plastics.
- Medical Devices: In the medical field, custom implants, orthopaedic devices, and prosthetics are made with biodegradable composites using FDM technology. Medical wastes and environmental impacts are reduced significantly.
- Construction: Construction materials like insulation panels, formwork, and structural elements are fabricated using FDM technology. It helps to create lightweight structures as material usage is less, which also considers sustainability.

1.4.2 Other Fabrication Techniques for Green Composites

Apart from FDM, there are several other fabrication methods that are utilized to manufacture green composites. These conventional methods involve compression moulding, injection moulding, RTM, and vacuum infusion, among others. Each method has its benefits and is applicable to various types of green composite materials and applications.

1.4.2.1 Compression Moulding

Compression moulding is the most common technique to make green composites, particularly for natural fibre-reinforced polymers. A preheated mixture of polymer and fibre (termed as a "charge") is filled in a mould cavity, closed, and treated with heat and pressure. The polymer is melted, and it flows into the mould and coats the natural fibres. The compression moulding produces composites with improved mechanical properties since there is high pressure applied during moulding. Compression moulding can be effective and be applied in fabricating large volumes of components within a short period and hence can be used for mass production. Compression moulding is also very cheap, especially where high-volume production is involved.

1.4.2.2 Injection Moulding

Injection moulding is another highly visible manufacturing technique for green composites. For this process, a biopolymer blend containing natural fibres is melted and pressed into a mould cavity under high pressure. The material fills the mould, and after cooling, the mould is opened to produce the final part. Injection moulding enables the fabrication of highly accurate, detailed parts with close tolerances. This is a highly automated process, and it is well-suited to mass production with repeat quality. Injection moulding saves material, which is a significant factor when working with bio-based and natural fibre composites.

1.4.2.3 Resin Transfer Moulding

RTM is an encapsulated moulding process wherein the dry natural fibres used for reinforcing are introduced inside the mould, and the biodegradable liquid resin is filled with pressure in the cavity of the mould. Resin will absorb inside the fibre, and when fully cured, remove the part-composite from the mould. Parts developed from RTM come out smooth over their surfaces, so they can find good usage within the aesthetical components.

The process can make large, complex, and high-performance parts with little material loss. As it is a closed-mould process, RTM produces fewer volatile organic compounds; thus, it is more environmentally friendly.

1.5 APPLICATIONS OF GREEN COMPOSITES

Green composites, consisting of natural fibres and biodegradable resins, have become increasingly popular with their eco-friendly character, sustainability, and potential to decrease dependency on non-renewable resources. Green composites provide a suitable alternative to conventional petroleum-based composites, making them more and more pertinent to industries that want to enhance their environmental impact. With industries and researchers on the lookout for new materials and technologies, the uses of green composites are being extended across a wide range of industries. This section delves into the wide variety of uses of green composites, mentioning their application in automotive, aerospace, construction, consumer goods, packaging, medical, and sports industries.

1.5.1 Automotive Industry

The automobile industry has been among the pioneering ones in the implementation of green composites, led by the pressure to reduce vehicle weights that optimize fuel efficiency while cutting greenhouse emissions. Natural fibre-based composites have been increasingly gaining traction with the push for manufacturers to switch to non-traditional materials from steel, aluminium, and glass fibre-reinforced plastics. One of the main uses of green composites in the automobile industry is in manufacturing interior parts like door panels, dashboards, seatbacks, and headliners. The parts are normally manufactured from natural fibres, reinforced with biodegradable or thermoplastic resins. Green composites are lighter in weight compared to conventional materials, and this assists in lowering the weight of the vehicle. Fuel efficiency is consequently enhanced, as well as reducing emissions. Natural fibres are also less costly than synthetic fibres, leading to material and manufacturing cost savings. Natural fibres and biodegradable resins ensure that automobile parts leave a smaller ecological footprint during their life cycle, from manufacturing through disposal. Besides interior parts, green composites are also used in structural parts of vehicles, including the production of underbody panels, trunk liners, and outer body panels. These parts utilize the mechanical stiffness and strength provided by natural fibre-reinforced composites. Although green composites provide weight savings and sustainability advantages, their mechanical properties, such as impact resistance and life, may require improvements to match conventional materials in severe applications.

1.5.2 Aerospace Industry

The aerospace sector, with its high demands in terms of strength, durability, and weight saving, has also looked at the application of green composites. While natural fibre-reinforced composite usage in aerospace is still in its infancy relative to automotive, interest in their usage for non-structural parts and interior components is increasing. Green composites are being utilized in the aerospace industry for the manufacture of interior cabin parts like seats, overhead stowage bins, tray tables, and wall panels. Such composites are usually processed from natural fibres such as flax, jute, and hemp with biodegradable resins.

Lightening an aircraft is one of the major factors in achieving maximum fuel efficiency and reducing operating costs. Green composites lighten aircraft without compromising their necessary strength and stiffness. The use of renewable fibres and biodegradable resins complements the aerospace industry's sustainability goals in the sense that it reduces the environmental impact of aircraft interiors. Though green composites have found limited application in aerospace structural systems today, efforts are ongoing to enhance their mechanical properties and resistance to fire in order to address the stringent requirements of performance and safety necessary in aerospace engineering. With the elimination of these limitations, the aerospace industry can anticipate a broader potential for green composites.

1.5.3 Construction Industry

The construction industry has traditionally relied on the utilization of materials such as steel, concrete, and wood, which all have significant environmental effects through the consumption of energy and resources. Green composites offer an effective alternative approach to the manufacture of sustainable, eco-friendly construction materials. Green composites may be used in all forms of application, varying from structural components to insulation and interior finishes. Green composites are gaining popularity in building panel production, cladding, and roofing panels. Composites are commonly reinforced with natural fibres like flax, jute, and bamboo, possessing high mechanical strength and weather resistance.

Natural fibre composites possess high resistance to temperature variation, ultra violet (UV) light, and moisture, hence find their application in exterior applications. Natural fibres have properties inherent in them, and therefore, green composites prove efficient in giving thermal insulation, thereby increasing the energy efficiency of the building. Green composites can be designed to reproduce the visual look of natural products such as wood or stone, thus providing aesthetic freedom without sacrificing sustainability. Green composites made from natural fibres are being researched as insulation materials due to their better thermal and acoustic performances. Hemp, flax, and coconut fibres are utilized for making insulation panels and mats that provide suitable heat and sound insulation in domestic and commercial buildings.

1.5.4 Consumer Goods

Green composites are becoming more common in a range of consumer goods due to the rising need of consumers for sustainability and green options instead of traditional plastics. Primary household items and personal care products are just a few examples of what consumers can purchase that are made from green composites. Natural fibre-reinforced green composites are applied for the production of furniture parts, for instance, seat frames, tabletops, and cabinets. They provide the necessary mechanical properties for durable pieces of furniture, while also serving as an eco-friendly substitute for plastic and particleboard headboards. Products that make up the frame of green composite furniture can be manufactured from renewable fibres and biodegradable resin, enabling the products to be green and compostable after usage. Green composite furniture is also more lightweight compared to traditional material furniture, thus facilitating easy transportation and erection of the furniture. Green composites offer a good sustainable alternative to sectors that utilize large amounts of plastic and, consequently, produce the most waste pollution. Now, natural fibres like bamboo, jute, and sugarcane fibres are used to create biodegradable packaging that can replace plastic packaging. The use of natural materials such as green composites helps to reduce the amount of waste plastic that is deposited at landfill sites and ocean edges, as they are biodegradable. Green composites made from natural fibre constituents are also more affordable than the traditional plastic materials, so they are much more beneficial for a business striving to save money and promote an eco-friendlier environment.

1.5.5 Medical Applications

The window of opportunity presented by these green composites fosters growth in the healthcare sector, which includes the usage in the making of bio-degradable green composites for medical devices, implants, and other health-care tools. Some green composites are best suited for temporary implants or single-use medical devices as they are biocompatible and biodegradable. Bone screws, plates, and meshes for fracture fixation, along with tissue engineering, are now made from natural fibres and PLA biodegradable resins.

These implants are designed to be bioactive and biodegradable with time so that a second surgery to remove the implant is not necessary.

Biodegradable green composites do not provoke inflammation, rejection, or other complications as they are non-toxic and biocompatible. Not needing to undergo a second surgery greatly reduces patient recovery time and the costs associated with the procedure, as the implants are meant to dissolve nat-urally within the body. Many other disposable products, like syringes, gloves, and surgical tools, are a primary source of waste in the healthcare sector. With the aid of green composites, single-use plastic medical instruments and devices can be made biodegradable, thus decreasing medical waste. Made from biodegradable green composites, these disposable medical products minimize the disposable. Green composite materials are very cost-efficient, making it economically feasible to produce.

1.5.6 Sports and Recreation

Another emerging area of application for green composites is the sports and recreation industry, for example, sporting equipment and outdoor sporting goods. Green composites are used as a substitute for more traditional materi-als and perform exceptionally well.

Green composites are currently being utilized to manufacture bicycles, tennis rackets, surfboards, and even golf clubs. The equipment is produced by reinforcing natural flax, hemp, or bamboo fibres with biodegradable resins to create lightweight, high-performance equipment that meets the demands of the sports enthusiasts and athletes. Composites used in the manufacture of sporting gear are lightweight yet rigid enough to make them suitable for high-performance sports. Biodegradable and renewable materials are used in construction of sports gear because the environmental impact is less during destruction.

Green composites are also used in production of camping, hiking, and also other sports gear. Tents, backpacks, sleeping bags, and other out-door camping gear can be lightweight and durable due to the use of natural fibre-based green composites. Green composites posses the required strength and endurance for outdoor applications even in extreme adverse weather conditions. The inclusion of biodegradable polymers in outdoor gear can further improve sustainability and ease disposal. The use of biodegradable materials in outdoor gear helps reduce the environmental footprint of outdoor enthusiasts, contributing to conservation efforts.

1.5.7 Packaging Industry

The biggest users of plastics are found in the packaging industry and there is a potential to decrease the negative impact that plastic waste has if these industries switch to green composites. There is a shift in the acceptance of novel biodegradable packaging materials among industries concerned about their plastic footprint. Green composites are being explored for developing new solutions for food packaging that are sustainable, biodegradable, and compostable. Fibres from bamboo, rice husk, and sugarcane are used with biodegradable resins to produce eco-friendly packaging that does not pollute even after being disposed of. Green-composited Phade® packaging is significantly more environmentally friendly than conventional plastic since they take an extremely shorter time to decompose, contributing positively to the reduction of landfill waste. Food packaging green composites are safe for the environment, non-toxic, and quite conversely beneficial for the end users, making them the ideal option for food industry practitioners who care for the environment.

Green composites are being applied in other consumer goods other than foods such as electronics, cosmetics, and household items. Studies show that these companies tend to prefer sustainable packaging materials to fulfil the demand for eco-friendly products from consumers.

1.5.8 Marine Industry

Glass fibre composites are now replaced with green composites particularly in making boat hulls to increase sustainability. The fibres that are made naturally are used as an alternative for fibreglass composites as they are used in boat hulls and decks. These green composites consist of jute, flax, and hemp and also have required strength and water resisting capacity. Green composites present a sustainable approach to boat manufacturing, lessening the environmental impact associated with fibreglass disposal at the end of a vessel's life. Natural fibres and biodegradable resins can be designed to withstand the tough conditions of marine environments, including saltwater and UV exposure.

1.6 IMPORTANCE OF GREEN COMPOSITES

Green composites, made up of natural fibres and biodegradable resins, had become more important as they address the cruellest point of the environment impact in the present time. The green composites have become an alternative to composites made up of petrol by minimizing the waste. This tells the multiple importance of the green composites, innovation, and social benefits.

1.6.1 Environmental Benefits

Probably, one of the main reasons for the growing importance of green composites is their environmental positivity. Traditional composite materials were usually made from non-renewable resources such as petroleum or any related products, which only contribute to pollution, wastage of energy, and carbon emissions. Green composites, by virtue of being developed on sustainability principles, intrinsically possess various environmental benefits.

1.6.1.1 Low Carbon Footprint

The conventional composites are manufactured using energy-intensive processes and high-carbon-emitting raw material processing. Green composites, however, are manufactured using natural fibres such as flax, jute, bamboo, and hemp, which consume low energy and are largely composed of renewable agricultural resources. These bio-composite materials function as carbon sinks during growth, capturing atmospheric CO_2 and minimizing the total carbon footprint of the end product. Further, biodegradable resins in green composites degrade naturally upon completion of their life cycle, unlike synthetic polymers, which last for centuries in the environment.

1.6.1.2 Biodegradability and Reduced Waste

Green composites are the answer to plastic waste, which has emerged over time as the most worrying environmental challenge. Traditionally, composites made from synthetic fibres and resins cannot be said to be decomposable; rather, they end up as litter in landfills and the sea, thus portraying long-term environmental degradation. Green composites, on the other hand, are sure to be biodegradable naturally, thus decomposing without any residual harm.

This makes them the best option for use in situations where material disposal is a paramount factor, for example, packaging, medical devices, and disposable consumer products.

1.6.1.3 Sustainable and Renewable Resources

Unlike traditional composites, whose sources are finite fossil fuels, green composites are derived from renewable resources. Green composites have the advantage of natural fibres coming from plants that can be harvested and returned for regrowth within a year. In addition, most of the crops employed in the production of these fibres, e.g., jute, bamboo, and hemp, use very little water, pesticides, and fertilizers, making them environmentally friendly compared to synthetic fibres like glass and carbon, which use high energy and raw materials in their manufacture.

1.6.2 Economic Advantages

Apart from their contribution to environmental sustainability, green composites are economically advantageous. With industries being pressured to embrace sustainable production and regulatory compliance, the economic value of green composites is increasingly evident. Their application can result in savings on production, material procurement, and disposal costs, as well as expansion of new business opportunities for sustainable products.

1.6.2.1 Cost-Effective Production

One of the most important economic advantages of green composites is the opportunity for cheaper production. Natural fibres, including jute, flax, and coir, are less expensive than glass and carbon, which are synthetic fibres. They are readily available in quantities, especially where the agriculture sector is strong, and need lesser resources to be processed. Besides that, the processes of manufacture for green composites tend to be lower in energy inputs than for conventional composites as well, hence adding to their cost-effectiveness. Further, governments are providing incentives and subsidies for sustainable material development and utilization, adding to green composites' competitiveness for manufacturers financially. These guidelines are put into effect so that the materials will be reused, recycled, or decompose once their purpose is over. Growing demand from consumers for environmentally friendly products has created new business opportunities for green composites. With increasing consciousness about the environmental impact of traditional materials, consumers now want to make a shift towards products that appeal to the values of sustainability and responsibility in consumption patterns. This has, in turn, increased the demand for green composite materials, particularly in industries such as auto, packaging, consumer durables, and infrastructure. Businesses leveraging green composites as part of their production mechanism can improve brand reputation by appearing environmentally conscious. This not only assists in terms of a bigger market share but also creates customer loyalty, as consumers tend to patronize brands that share their environmental concerns.

1.6.3 Versatility and Wide Range of Applications

Green composites are versatile; therefore they can be used in various industries. Their lightweight and mechanical strength properties make them suitable for use in automotive and aerospace applications and in consumer goods and medical products.

1.6.3.1 Automotive Industry

Perhaps the most far-reaching place where green composites have impacted is the automotive sector. The need for lighter materials to enhance fuel economy and lower emissions has prompted car manufacturers to seek alternatives to conventional materials such as steel and plastic. Green composites are an alternative, since they offer the strength and stiffness needed but with much lighter weights. Natural fibre-reinforced composites are finding more uses for interior parts such as door panels, seatbacks, and dashboard structures.

1.6.3.2 Construction Sector

In the building construction sector, green composites find applications in structural elements, insulation, and cladding. Their thermal and sound insulation, moisture, and UV degradation resistance make them suitable for building panels, roofing, and other construction applications. Furthermore, the application of green composites in building construction is also aligned with the sector's increasing emphasis on sustainable construction practices, such as energy-efficient and environmentally friendly designs.

1.6.3.3 Medical and Healthcare

The degradability and biocompatibility of green composites make them a good choice for medical use such as biodegradable implants, wound dressing, and disposable medical devices. For instance, composite materials based on PLA are being researched to be used in the creation of degradable plates and screws to set broken bones. These implants degrade gradually in the body after a period of time, thereby eliminating the requirement to remove such implants via another surgery. This lowers the expense of healthcare but also enhances patients' results.

1.6.3.4 Consumer Products and Packaging

Green composites in the consumer goods sector are applied in the production of furniture, enclosures for electronics, and personal care products. Their strength and light weight make them an excellent material for daily use. Also, in packaging, green composites provide a biodegradable option to plastic packaging, which is the problem of plastic waste globally. Biodegradable resin and natural fibre packaging can be composted, eliminating the footprint of disposal on the environment.

1.6.4 Contribution to Sustainable Development

Green composites are the main component in sustainable development, which aligns with the United Nations' Sustainable Development Goals. They contribute in promoting sustainability, reducing waste, and also in efficient consumption and production.

1.6.4.1 Resource Efficiency

Utilizing natural fibres obtained from renewable sources, green composites aid in the promotion of resource efficiency. The crops that yield such fibres, like jute, bamboo, and flax, are extremely renewable and can be planted with practically no harmful effect on the environment. The added advantage of biodegradable resins creating greener composites is that they can be formulated with renewable feedstocks like corn starch or sugarcane. They are therefore less reliant on limited fossil fuels and reduce the total environmental impact of material production.

Green composites follow the principles of the circular economy by encouraging the utilization of materials that are recyclable or biodegradable. Unlike traditional composites, which are difficult to recycle due to their complexity, green composites can degrade naturally or be recycled and reused as new products. This avoids final-life waste and minimizes the environmental impact of disposal. For packaging and other industries, the use of green composites provides a solution to the global issue of plastic waste. Green composite biodegradable packaging in natural environments eliminates plastic trash from landfills and oceans. This results in healthier environments and reduces the burden on waste management.

1.6.5 Technological Advancements and Innovation

The evolution of green composites has driven tremendous progress in material science and manufacturing technologies. With new natural fibres, biodegradable resins being discovered by researchers, and new fabrication methods, the usage scope of green composites is expanding. Not only does the evolution improve the performance of green composites, but it also creates new uses for them in many industries. Researchers continue to look for newer natural fibres with improved mechanical properties to enhance the performance of green composites. For example, improvement in treatment and processing of natural fibres like bamboo, coir, and kenaf has improved their stiffness, strength, and durability. These natural fibres with high strength are enabling green composites to compete with synthetic composites in high-performance applications, e.g., automotive and aircraft components.

1.6.5.1 Biodegradable Resin Advancements

Evolution of biodegradable resins has also been a focus area for green composites. PLA, PHAs, and starch resins are a few polymers that are best suited for applications owing to their biodegradability and renewability characteristics. Further, the development of bio-based epoxy resins, which are stronger and resistant to environmental conditions, is increasing the use base of green composites in high-performance applications.

1.6.5.2 Reduction of Reliance on Fossil Fuels

Among the most significant aspects of the role of green composites is their role to reduce fossil fuel dependency. Traditional composite materials, such as those reinforced with glass or carbon fibres, are usually produced through energy-consuming processes involving petroleum-based resins. Green composites, however, utilize biodegradable resins from renewable materials such as corn starch, sugarcane, and vegetable oils. These bio-resins reduce the consumption of fossil fuels during production and assist in reducing greenhouse gas emissions. Furthermore, natural fibres used in green composites can be cultivated with minimal energy input as opposed to synthetic fibres, which require high levels of energy to produce. This shift from fossil fuel-based materials to renewable materials reduces the impact of composite production on the environment and helps the world transition towards a cleaner energy infrastructure.

1.7 CONCLUSION

Green composites have emerged as a pivotal technology in sustainable materials science, both tackling environmental and economic issues and driving the limits of technological progress. This chapter has provided an in-depth overview of green composites, discussing fundamental aspects ranging from their history to materials, manufacturing methods, applications, and general significance within the framework of global sustainability.

Section 1.2 chronicled the development of green composites from the initial incorporation of natural fibres to the recent research on biodegradable resins. It traced the landmarks through the evolutional process that included the escalating realization of the eco-footprint of conventional composites, thereby initiating the pursuit for greener substitutes. It symbolizes how much more concern for eco-friendliness gained attention as awareness of environmental care increased among industries.

Section 1.3 gave a detailed account of the natural fibres and biodegradable resins that are the cornerstone of green composites. Natural fibres like jute, bamboo, coir, and cane bring not just renewability but also mechanical properties as well as biodegradability, making them perfect replacements for synthetic fibres. Similarly, biodegradable resins like PLA and PHA offer answers that aim towards reducing waste and minimizing fossil fuel use. Their application is illustrative of the natural sustainability and eco-friendly nature of green composites. In Section 1.4, several processes, including FDM, compression moulding, and injection moulding, were outlined.

These processes play a critical role in defining the quality and performance of green composites in different applications. FDM, however, presents novel possibilities for the fabrication of complex and tailored parts with sustainability, while other methods, such as compression and injection moulding, have been optimized for decades to incorporate green materials into industrial manufacturing. The progressive optimization of these fabrication processes allows green composites to match traditional composites in performance while providing environmental advantages. Section 1.5 analyzed the various sectors in which such materials have established themselves in earnest. Ranging from the auto and aviation industries to medical implants, construction, packaging, and consumer products, green composites have proven versatile and efficient at mitigating their effects on the environment.

Natural fibre composites have been embraced by the automotive industry to reduce the weight and emissions of cars, and the construction industry has embraced green composites as a material for green buildings. Medical uses, including biodegradable implants, illustrate the ability of green composites to improve patient health while minimizing waste in the environment. These applications are clear signs of growing importance and utility of green composites in modern society. Finally, Section 1.6 concluded the overall worth that these materials bring to industry and the environment. Green composites not only offer a solution to the pressing issues of climate change, resource depletion, and waste minimization, but also offer economic benefits through lower production costs and the creation of new business opportunities for green products. Their contribution cannot be overvalued because they facilitate world agendas aimed at mitigating carbon emissions, preserving resources, and propagating the circular economy.

Shift to fossil fuel-replacing non-renewables, specifically towards biodegradable materials or biological ones, marks an important landmark to achieving greener and a sustainable tomorrow. The issues covered under this introductory chapter, ranging from the history of green composites to materials, manufacturing process, applications, and significance, serve to underscore the need for such materials in the future. Increasing use of green composites across numerous industries is a reflection of a broader shift towards sustainability stimulated by regulatory imperatives as much as by demands from consumers for green products. As technology and science advance the limits of the properties and applications of green composites, their contribution to the creation of a sustainable and environmentally conscious future will grow. This chapter establishes the basis for a more detailed discussion of the specific challenges and opportunities of green composites, as outlined in the subsequent chapters of this book.

REFERENCES

1. Abdur Rahman, M., Serajul Haque, Muthu Manokar Athikesavan, and Mohamed Bak Kamaludeen. "A review of environmental friendly green composites: Production methods, current progresses, and challenges." *Environmental Science and Pollution Research* 30, no. 7 (2023): 16905–16929.
2. Naik, Nithesh, B. Shivamurthy, B. H. S. Thimmappa, Gautam Jaladi, Kaustubh Samanth, and Nagaraj Shetty. "Recent advances in green composites and their applications." *Engineered Science* 21, no. 5 (2022): 779.

3. Kuram, Emel. "Advances in development of green composites based on natural fibers: A review." *Emergent Materials* 5, no. 3 (2022): 811–831.

4. Haris, Nur Izzah Nabilah, Syeed SaifulAzry Osman Al Edrus, Norfaryanti Kamaruddin, Nurliyana Abdul Raof, A. W. Noraida, and Mohd Hafizz Wondi. "Environmental and economic impacts of green panels." *In Green Lignocellulosic-Based Panels: Manufacturing, Characterization and Applications*, edited by Syeed SaifulAzry Osman Al Edrus, Seng Hua Lee, Juliana Abdul Halip, Norfaryanti Kamaruddin, Nur Izzah Nabilah Haris, and Paridah Md Tahi, pp. 191–209. Springer Nature Singapore, 2025.

5. Stavropoulos, Panagiotis, and Vasiliki Christina Panagiotopoulou. "Carbon footprint of manufacturing processes: Conventional vs. non-conventional." *Processes* 10, no. 9 (2022): 1858.

6. Rajendran, Sundarakannan, Ali Al-Samydai, Geetha Palani, Herri Trilaksana, Thanikodi Sathish, Jayant Giri, R. Saravanan, J. Isaac JoshuaRamesh Lalvani, and Faouzi Nasri. "Replacement of petroleum based products with plant-based materials, green and sustainable energy—A review." *Engineering Reports* 7, no. 4 (2025): e70108.

7. Maiti, Saptarshi, Md Rashedul Islam, Mohammad Abbas Uddin, Shaila Afroj, Stephen J. Eichhorn, and Nazmul Karim. "Sustainable fiber-reinforced composites: A review." *Advanced Sustainable Systems* 6, no. 11 (2022): 2200258.

8. Rajeshkumar, G., S. Arvindh Seshadri, G. L. Devnani, M. R. Sanjay, Suchart Siengchin, J. Prakash Maran, Naif Abdullah Al-Dhabi, et al. "Environment friendly, renewable and sustainable poly lactic acid (PLA) based natural fiber reinforced composites—A comprehensive review." *Journal of Cleaner Production* 310 (2021): 127483.

9. Khatri, Hardik, Jesuarockiam Naveen, M. Jawaid, K. Jayakrishna, M. N. F. Norrrahim, and Ahmad Rashedi. "Potential of natural fiber based polymeric composites for cleaner automotive component production—A comprehensive review." *Journal of Materials Research and Technology* 25 (2023): 1086–1104.

10. Islam, Tarikul, Mehedi Hasan Chaion, M. Abdul Jalil, Abu Sayed Rafi, Faujia Mushtari, Avik Kumar Dhar, and Shahin Hossain. "Advancements and challenges in natural fiber-reinforced hybrid composites: A comprehensive review." *SPE Polymers* 5, no. 4 (2024): 481–506.

11. Mohammed, Mohammed, Jawad K. Oleiwi, Aeshah M. Mohammed, Anwar Ja'afar Mohamad Jawad, Azlin F. Osman, Tijjani Adam, Bashir O. Betar, and Subash CB Gopinath. "A review on the advancement of renewable natural fiber hybrid composites: Prospects, challenges, and industrial applications." *Journal of Renewable Materials* 12, no. 7 (2024): 1237–1290.

12. Nagaraja, K. C., Attel Manjunath, K. G. Pranesh, G. Rajeshkumar, M. G. Raju, and Shrimukhi G. Shastry. "Impact of manufacturing methods on the mechanical characteristics of polymer matrix composites with glass and epoxy fiber reinforcement." *Journal of the Institution of Engineers (India): Series D* (2024): 1–7.

13. Asmawi, Nazrin, Tze Mei Kuan, and Vasi Uddin Siddiqui. "Manufacturing of fiber polymer composite materials." In *Aerospace Materials*, edited by Mohamed Thariq Hameed Sultan, Marimuthu Uthayakumar, Kinga Korniejenko, Peter Madindwa Mashinini, Muhammad Imran Najeeb, and Renga Rao Krishnamoorth, pp. 333–347. Elsevier, 2025.

14. Ghosh, Suman Kumar, and Narayan Chandra Das. "Recent advances in nanoclay- and graphene-based thermoplastic nanocomposites for packaging applications." *Packaging Technology and Science* 37, no. 6 (2024): 503–531.

15. Ramesh, Manickam, and C. Deepa. "Processing and properties of jute (*Corchorus olitorius* L.) fibres and their sustainable composite materials: A review." *Journal of Materials Chemistry A* 12, no. 4 (2024): 1923–1997.

16. Ray, Bankim Chandra, Rajesh Kumar Prusty, Dinesh Kumar Rathore, Sohan Kumar Ghosh, and Abhijeet Anand. "Micro and nanophased polymeric com- posites: Durability assessment in engineering applications." (2024). Woodhead publishing.

17. Yu, Haixia, Yahui Zhang, Jingpeng Li, and Fei Rao. "Structural characteristics and physicomechanical properties of bamboo scrimber composite during natu- ral weathering." *Surfaces and Interfaces* 51 (2024): 104714.

18. Kumar, Rahul, Faladrum Sharma, Sumit Bhowmik, and PMG Bashir Asdaque. "Recent trends in coconut coir fibre-reinforced composite material." In *Composites*, edited ByVijay Kumar Singh, Nishant Kumar Singh, Yashvir Singh, pp. 101–125. CRC Press, (2024).

19. Anerao, Prashant, Atul Kulkarni, and Yashwant Munde. "A review on explora- tion of the mechanical characteristics of 3D-printed biocomposites fabricated by fused deposition modelling (FDM)." *Rapid Prototyping Journal* 30, no. 3 (2024): 430–440.

Harnessing the Power of Statistics in Green Composite Modelling

2

2.1 INTRODUCTION

The global pursuit of sustainable development, climate resilience, and environmental stewardship has prompted the scientific community and industrial sectors to rethink traditional materials and manufacturing processes. One of the key strategies in this transition is the shift towards eco-conscious materials that reduce environmental burdens across their life cycle. Among these alternatives, green composites have emerged as a compelling class of materials. These are typically made from natural fibres such as jute, flax, kenaf, or sisal, and biodegradable or bio-based polymers such as polylactic acid (PLA), polyhydroxybutyrate (PHB), and thermoplastic starch, which serve as matrix materials. Green composites offer the dual advantage of reduced dependency on fossil resources and enhanced end-of-life biodegradability, making them suitable candidates for the next generation of sustainable engineering applications [1,2].

However, despite their environmental advantages, green composites face several scientific and technological barriers. The mechanical performance of these materials is often inferior to that of their synthetic counterparts, due to variability in fibre morphology, poor fibre-matrix interfacial bonding, and moisture sensitivity. In response to these challenges, researchers have increasingly turned to advanced manufacturing technologies, particularly fused deposition modelling (FDM), to explore novel fabrication pathways and achieve tailored mechanical properties. FDM, a subset of additive manufacturing (AM), offers unique advantages for green composite fabrication, such as low material wastage, rapid prototyping, flexibility in geometric design, and the ability to process a wide range of thermoplastic materials [3,4].

2.1.1 The Promise and Complexity of Green Composite Fabrication via FDM

The integration of green composites into FDM presents both opportunities and complexities. On one hand, FDM can be instrumental in developing functional parts from biodegradable feedstock, supporting eco-design principles and waste reduction strategies. On the other hand, the process involves multiple interacting parameters, such as layer thickness, nozzle temperature, infill density, raster orientation, and printing speed, that significantly influence the final properties of the fabricated part. The addition of natural fibres further complicates the scenario by introducing heterogeneities, affecting thermal conductivity, viscosity, and extrusion flow behaviour during printing [5,6]. Such complexity necessitates a scientifically grounded, data-centric approach to understand and optimize the FDM process for green composites. In the absence of robust modelling frameworks, designers and engineers are forced to rely on iterative trial-and-error practices, which are time-consuming, resource-intensive, and may not lead to generalized or scalable conclusions. It is in this context that statistical modelling and analysis become indispensable.

2.1.2 Role of Statistical Modelling in Green Composite Research

Statistical modelling provides a structured framework to analyze, interpret, and predict how process parameters affect material properties. Unlike purely empirical methods, statistical tools facilitate the systematic exploration of multi-variable, non-linear relationships and help isolate the effect of individual parameters or their interactions [7]. Among the most applied statistical techniques in this domain are:

- **Design of Experiments (DOE):** A strategy to plan experiments such that all combinations of factors are explored efficiently. It includes factorial designs, fractional factorial designs, and central composite designs (CCDs).
- **Response Surface Methodology (RSM):** A collection of mathematical and statistical techniques useful for modelling and analyzing problems in which a response is influenced by several variables. RSM helps in building predictive models and optimizing responses.
- **Analysis of Variance (ANOVA):** A tool used to identify statistically significant factors and interactions.
- **Regression Analysis:** Linear, non-linear, or polynomial regression is used to fit experimental data to predictive models.
- **Taguchi Methods:** A robust design strategy aimed at improving quality and performance through parameter optimization while minimizing the effects of noise.
- **Machine Learning (ML) Techniques:** More recently, models like artificial neural networks (ANNs), support vector machines (SVMs), and random forests have been employed for modelling complex relationships in green composite manufacturing.

The application of these tools allows researchers to identify optimal processing windows, determine the sensitivity of properties to various inputs, and predict the behaviour of the composite under untested conditions.

2.1.3 Advancing Process Understanding through Statistical Insights

In the context of green composite modelling, statistical approaches serve multiple functions:

1. **Parameter Screening and Selection:** Not all process parameters have a significant effect on output responses. DOE helps identify critical parameters that warrant close control, allowing for more efficient use of resources.

2. **Optimization of Process Conditions:** By constructing response surfaces and analysing interaction plots, RSM enables multi-objective optimization. For example, one can simultaneously optimize for tensile strength, surface finish, and biodegradability.

3. **Understanding Material Behaviour:** Statistical models reveal how natural fibre content influences thermal shrinkage, how infill density affects porosity, or how raster orientation impacts anisotropy in mechanical strength.

4. **Predictive Modelling:** Once validated, regression or ML models can serve as surrogates for physical experimentation, thus accelerating product development cycles.

5. **Quality Assurance and Robustness:** Statistical control charts, capability indices, and robust design methods ensure repeatable part quality under variable environmental or operational conditions.

6. **Sustainability Assessment:** Statistical techniques can also be integrated into life cycle analysis (LCA) models to understand how changes in process variables affect energy consumption, carbon footprint, and material usage [8,9].

2.1.4 Interdisciplinary Nature of Statistical Green Composite Research

The domain lies at the intersection of multiple disciplines: materials science, mechanical engineering, manufacturing systems, and statistical analytics. As such, the modelling and analysis of green composites fabricated via FDM cannot be addressed in silos. A successful strategy demands:

- **Material Expertise:** Understanding the physical, chemical, and mechanical behaviour of natural fibres and biopolymers.
- **Process Understanding:** Deep knowledge of AM workflows and the physical interactions occurring during deposition.
- **Data Literacy:** Ability to collect, clean, and analyze data, along with a sound understanding of statistical principles and software tools.
- **Sustainability Lens:** Evaluation of process and material choices in terms of environmental and economic impacts.

This necessitates a collaborative research framework, where domain experts work in tandem with data scientists to co-develop models that are both accurate and interpretable.

2.1.5 From Statistical Analysis to Intelligent Manufacturing

The increasing digitization of manufacturing systems is paving the way for Industry 4.0 paradigms to be applied to green composites [10,11]. With the integration of sensors, real-time data acquisition systems, and Internet of Things-enabled devices, statistical models can now operate in a closed-loop control fashion. In such systems, real-time feedback from process parameters or part quality can be used to adjust FDM settings dynamically, thereby achieving adaptive manufacturing. Moreover, the emerging concept of the digital twin, a virtual replica of the physical process, relies heavily on accurate predictive models developed through statistical and ML techniques. In the future, digital twins for green composite fabrication could help simulate a wide range of materials, fibre orientations, and process conditions, thereby informing design decisions without physical experimentation.

2.1.6 Challenges in Statistical Modelling of Green Composites

While the benefits of statistical modelling are numerous, the journey is not without challenges:

- **Data Variability:** Natural fibres are inherently variable due to their biological origin. This introduces noise and uncertainty into datasets, which must be accounted for in model design.
- **Non-Linearities and Interactions:** Many processes involved in FDM are non-linear and exhibit strong interaction effects, making linear models inadequate. Advanced modelling techniques or transformations may be required.
- **Overfitting in ML Models:** Complex models can often fit the training data too closely, resulting in poor generalization. Careful cross-validation and model selection techniques are necessary.
- **Experimental Costs:** Running full factorial designs or complex response surface models may require many experiments, which can be costly in terms of materials, time, and equipment.
- **Interpretability:** As more ML techniques are applied, the interpretability of the models becomes an issue, particularly for regulatory or industrial adoption where explain ability is critical.

Despite these hurdles, advances in computational power, open-source statistical tools, and increasing awareness of data-driven methodologies are steadily overcoming these limitations.

2.1.7 Chapter Overview and Structure

This chapter sets the stage for a deep dive into how statistical tools can be strategically deployed to advance the modelling and optimization of green composites manufactured via FDM. The subsequent sections will explore:

- An overview of statistical tools applicable to FDM-based green composite modelling.
- Case studies illustrating the optimization of process parameters using RSM and Taguchi methods.
- Construction and validation of predictive models for mechanical and functional properties.
- Integration of statistical models into sustainable design frameworks and decision-support systems (DSSs).
- Limitations, best practices, and future research directions in statistical modelling of green composites.

2.2 OVERVIEW OF STATISTICAL TOOLS APPLICABLE TO FDM-BASED GREEN COMPOSITE MODELLING

The fabrication of green composites via FDM involves a complex interplay of material properties, process parameters, and environmental influences. These parameters significantly affect key quality indicators such as tensile strength, impact resistance, dimensional accuracy, surface finish, and biodegradability. Due to the inherent variability in natural fibres and the non-linear nature of FDM processes, statistical tools have become essential in understanding, modelling, and optimizing FDM-based green composite fabrication [12,13]. This section outlines the core statistical methodologies commonly employed in green composite research and highlights their significance in achieving data-driven, reproducible, and optimized outcomes. The statistical tools discussed include both classical techniques and modern data-driven approaches, each tailored for different stages of the modelling and analysis process.

2.2.1 Design of Experiments

DOE is a foundational statistical tool that enables systematic experimentation by varying multiple factors simultaneously. In the context of FDM, these factors could include nozzle temperature, layer height, print speed, infill density, fibre weight percentage, and build orientation. Key types of DOE used in green composite modelling:

- Full Factorial Designs: Explore all possible combinations of factors at different levels. Suitable for early stage exploratory studies where interactions are unknown.
- Fractional Factorial Designs: Reduce the number of experiments while preserving the ability to detect main effects and some interactions.
- Plackett–Burman Designs (PBDs): Useful for screening many variables to identify the most influential ones.
- CCD and Box–Behnken Design (BBD): Frequently used in RSM for fitting second-order polynomial models to capture non-linear behaviour.

DOE is a cornerstone in scientific experimentation and statistical modelling, enabling researchers to systematically investigate the relationship between input parameters (factors) and output responses. In the domain of green composite fabrication using FDM, DOE techniques play a pivotal role in uncovering hidden patterns, evaluating factor interactions, minimizing experimental trials, and enhancing process optimization. Natural fibre-reinforced composites exhibit high variability due to the heterogeneity of bio-based materials. When combined with the complex process dynamics of FDM, this makes DOE an indispensable approach. The most frequently employed DOE strategies in this context are full factorial designs, fractional factorial designs, PBDs, and CCD and BBD, particularly when applying RSM. Each of these DOE strategies serves a specific objective in the modelling life cycle, from screening to optimization.

2.2.1.1 Full Factorial Designs

The full factorial design [14] is the most exhaustive form of DOE. It involves studying every possible combination of factor levels, allowing for a complete exploration of all main effects and interaction effects.

Definition and Structure: For k factors, each at n levels, the total number of experimental runs is n^k. For instance, a three-factor, two-level design requires $2^3 = 8$ experimental runs.

2.2.1.1.1 Advantages
- Offers complete understanding of the system.
- Captures both main effects and interactions (including higher-order interactions).
- Useful when the experimental region is small and precision is critical.

Application in Green Composites: In early stage research involving FDM-printed biocomposites (e.g., PLA reinforced with jute or hemp), full factorial designs are often used to evaluate how fibre weight percentage, nozzle temperature, and printing speed affect mechanical properties such as tensile strength and modulus.

2.2.1.1.2 Limitations
- Rapidly grows in complexity with an increasing number of factors and levels (combinatorial explosion).
- Not ideal for resource-constrained environments or when working with expensive/rare materials.

2.2.1.2 Fractional Factorial Designs

When a full factorial design becomes impractical due to cost, time, or material constraints, the fractional factorial design offers a pragmatic alternative by selecting a carefully chosen subset (fraction) of the full design.

Definition and Structure: A fractional factorial design uses a fraction of the total runs in a full factorial design while still capturing the most critical effects (usually main effects and some two-way interactions).

Example: A half-fraction (1/2) of a 242^4 design requires only 8 runs instead of 16.

2.2.1.2.1 Advantages
- Reduces the number of experiments significantly.
- Efficient for screening purposes when the goal is to identify important factors.
- Captures main effects and low-order interactions, which often dominate in process control.

Application in Green Composites: Suppose a study examines six factors (layer height, bed temperature, fibre orientation, infill density, raster angle, and cooling rate). A full factorial design would require $2^6 = 64$ runs at two levels. A fractional factorial design can reduce this to 16 or 32 runs while still providing meaningful insight.

2.2.1.2.2 Considerations
- May result in confounding (aliasing) of interactions with each other.
- Selection of the correct resolution design (e.g., Resolution III, IV, or V) is crucial for accurate interpretation.

2.2.1.3 Plackett–Burman Designs

The PBD is a highly efficient screening design aimed at identifying the most influential factors among many with a minimum number of runs. It assumes that interactions between factors are negligible, focusing solely on estimating main effects.

Definition and Structure: PBDs are two-level designs based on multiples of four experimental runs (e.g., 8, 12, 16). Each factor is assigned to one column, and dummy variables may be used to estimate error.

2.2.1.3.1 Advantages
- Very economical and quick to implement.
- Ideal for initial screening in exploratory studies involving many variables.
- Helps eliminate insignificant factors before proceeding to detailed analysis.

Application in Green Composites: When dealing with ten or more processing variables (such as fibre size, matrix viscosity, drying time, print cooling, ambient humidity.), PBD is a rapid method to identify which few factors truly affect responses like interlayer adhesion, void content, or environmental degradation.

2.2.1.3.2 Limitations
- No interaction effects are modelled.
- Designed under the assumption that interactions are negligible (which may not always be true in FDM-based processes).
- Further analysis is needed to refine models after screening.

CCD and BBD: Both CCD and BBD are widely used in RSM to build accurate second-order (quadratic) models for optimization of process parameters. They allow the modelling of curvature (non-linear behaviour) and interactions between variables.

2.2.1.4 Central Composite Design

CCD augments a factorial or fractional factorial design with additional centre points and axial points to allow for curvature modelling.

2.2.1.4.1 Structure
- Factorial portion: Basic factorial design at low (−1) and high (+1) levels.
- Axial points (*α*): Points outside the factorial space to estimate curvature.
- Centre points: Replicated runs at the mid-level of all factors to estimate pure error and check for reproducibility.

2.2.1.4.2 Advantages
- Capable of estimating all quadratic effects.
- Can be tailored to fit within a practical or theoretical design space.
- Suitable for developing accurate predictive models.

Application in Green Composites: CCD has been applied to optimize printing temperature, fibre weight ratio, and build orientation to maximize impact strength and dimensional accuracy of PLA-linen fibre composites. The resulting regression models are used for sensitivity analysis and multi-objective optimization.

2.2.1.5 Box–Behnken Design

BBD is another popular second-order design that does not contain extreme factor combinations (unlike CCD), making it more efficient and safer in scenarios involving fragile or expensive materials.

2.2.1.5.1 Structure
- Uses midpoints of edges of the experimental space.
- Requires fewer runs than CCD for the same number of factors.
- Each factor is varied over three levels (low, medium, high), but does not include corner points of the cube.

2.2.1.5.2 Advantages
- Efficient in terms of number of runs.
- Avoids combinations of extreme factor levels, reducing the risk of failure or defects.
- Good for systems where non-linearity is expected but experimental space must be constrained.

Application in Green Composites: BBD is suited for optimizing eco-friendly FDM filaments where extreme combinations of fibre loading and temperature might cause nozzle clogging or thermal degradation. For example, optimizing cooling fan speed, layer height, and fibre content for maximizing flexural strength and surface finish of bamboo/PLA composites.

The choice of DOE technique (Table 2.1) in green composite modelling must align with the objective of the study, available resources, and the expected behaviour of the process. Whether it is early stage screening, detailed interaction modelling, or final process optimization, each design offers a unique set of advantages. With rising environmental concerns and increased adoption of natural fibres in AM, statistically designed experiments ensure that green composites are developed in a scientific, systematic, and resource-efficient manner. DOE techniques not only improve process understanding and product performance, but also contribute to sustainable material development by reducing experimental waste.

TABLE 2.1   Summary of DOE utility in green composites

DESIGN TYPE	PRIMARY USE	FACTOR LEVELS	INTERACTIONS MODELLED	STRENGTHS
Full factorial design	Comprehensive modelling	2+	All	Complete understanding of effects
Fractional factorial	Efficient exploration	2+	Some	Reduces experiments, still informative
Plackett–Burman design	Screening	2	None	Identifies key factors quickly
Central composite design	Curvature modelling and optimization	3+	All	Excellent for RSM and optimization
Box–Behnken design	Safe optimization in constrained space	3	All (second order)	Fewer runs, avoids extremes

Abbreviations: DOE, design of experiments; RSM, response surface methodology.

2.2.2 Response Surface Methodology

RSM is an advanced statistical and mathematical technique used for developing, improving, and optimizing processes. It is particularly well-suited for modelling and analyzing systems in which a response of interest is influenced by several input variables (also known as factors), with the goal of optimizing this response. RSM blends DOE with regression modelling and optimization algorithms to create empirical models that capture the behaviour of complex systems in a resource-efficient manner. In the context of FDM-based green composite fabrication, where natural fibre-reinforced thermoplastic polymers are used, RSM proves invaluable for understanding the effects of multiple process parameters (e.g., nozzle temperature, fibre content, layer thickness, and print speed) on crucial output responses such as mechanical strength, print accuracy, and material sustainability.

2.2.2.1 Goals of RSM

RSM serves several key purposes in scientific and engineering research:

2.2.2.1.1 Develop Empirical Models to Approximate True Response Surfaces

In real-world systems, especially those involving biodegradable or natural fibre composites, the mathematical representation of the process behaviour is unknown or too complex to be modelled analytically. RSM allows researchers to develop empirical (statistical) models that represent the relationship between inputs and outputs through regression equations. These models, often second-order polynomials, approximate the true response surface with sufficient accuracy in the vicinity of the optimal conditions. A general second-order RSM model is expressed as:

$$Y = \beta_0 + \sum \beta_i X_i + \sum \beta_{ii} X_i 2 + \sum \beta_{ij} X_i X_j + \varepsilon$$

where:

- Y = Response variable (e.g., tensile strength),
- X_i = Input variables (e.g., nozzle temperature, fibre content),
- β = Coefficients to be estimated,
- ε = Error term.

2.2.2.1.2 Identify Optimal Process Conditions

By fitting a response surface and analyzing its shape (e.g., using contour plots or 3D response plots), RSM helps identify optimal settings of process parameters that maximize or minimize the response variables. In the FDM context, this could mean finding the best combination of printing parameters to achieve maximum tensile strength, minimum warpage, or optimal surface finish.

2.2.2.1.3 Understand Factor Interactions and Response Curvature

Unlike simple linear models, RSM accounts for:

- Interaction effects: How two or more factors together influence the response.
- Curvature: Non-linearities in how a response changes across levels of a factor.

This enables deeper process understanding, critical when dealing with bio-based materials that show non-linear and interactive effects due to fibre-matrix compatibility, moisture absorption, and degradation.

2.2.2.2 RSM in FDM-Based Green Composite Fabrication

The FDM process, although versatile, involves a multitude of parameters, each contributing to the final part quality. When green composites are used (such as PLA reinforced with natural fibres like jute, hemp, kenaf, bamboo, or flax), the system becomes even more complex due to the anisotropic, heterogeneous, and biodegradable nature of materials. RSM serves as a robust tool to navigate and optimize such systems.

2.2.2.2.1 Input Parameters in FDM of Green Composites

Key process parameters (independent variables) that can be varied include:

- Nozzle temperature
- Bed temperature
- Print speed
- Layer height
- Fibre volume fraction
- Raster angle
- Infill density
- Cooling fan speed

2.2.2.2.2 Output Responses (Dependent Variables)

RSM allows for modelling and optimization of one or more of the following responses:

- Tensile, flexural, or impact strength
- Surface roughness
- Porosity or void fraction
- Dimensional accuracy
- Interlayer adhesion
- Strength-to-weight ratio
- Thermal stability
- Degradability and sustainability metrics

2.2.2.3 Implementation Steps of RSM in FDM Green Composite Research

2.2.2.3.1 Step 1: Experimental Design Selection

Common RSM-based designs include:

- CCD
- BBD

These designs are chosen for their ability to efficiently explore a design space with a minimum number of experimental runs, especially for second-order modelling.

2.2.2.3.2 Step 2: Conducting Experiments

Experiments are performed as per the design matrix, varying factors at coded levels (typically −1, 0, and +1).

2.2.2.3.3 Step 3: Model Building

A regression model is developed (usually quadratic) using the least squares method. ANOVA is performed to:

- Determine model adequacy.
- Assess the significance of each term.
- Identify lack-of-fit if any.

2.2.2.3.4 Step 4: Response Analysis

Contour plots and surface plots help visualize the influence of parameters. Statistical metrics such as R^2, adjusted R^2, predicted R^2, and p-values indicate the robustness of the model.

2.2.2.3.5 Step 5: Optimization
Optimization can be single-objective (e.g., maximizing tensile strength) or multi-objective (e.g., minimizing surface roughness while maximizing strength). The desirability function approach is commonly used for multi-response optimization.

2.2.2.3.6 Step 6: Validation
The predicted optimal settings are validated through confirmatory experiments, ensuring the model's accuracy in real-life scenarios.

2.2.2.4 Case Study: RSM in FDM-Based Green Composite Modelling

Consider a study aiming to optimize the FDM process for a PLA-hemp fibre composite. The objective is to maximize tensile strength and minimize surface roughness. Factors considered include:

- Nozzle temperature (°C)
- Layer height (mm)
- fibre content (% by weight)

Using CCD, the researchers perform 20 runs and develop a quadratic model. The analysis reveals:

- Significant interaction between nozzle temperature and fibre content
- A non-linear effect of layer height on tensile strength
- An optimal parameter combination of 205°C nozzle temperature, 0.15 mm layer height, and 8% fibre content

The confirmatory test at these settings shows a 7% increase in strength and a 15% reduction in surface roughness compared to baseline settings.

This case illustrates how RSM can help balance mechanical performance and surface quality in biodegradable FDM materials.

2.2.2.5 Advantages of Using RSM in FDM Green Composite Modelling

- Resource efficiency: Reduces the number of experimental trials needed.
- Enhanced process understanding: Reveals how parameters interact and influence each other.

- Robust optimization: Allows simultaneous optimization of multiple outputs.
- Predictive capability: Once validated, the model can be used to simulate responses for new settings.
- Scalability: The approach can be extended to new materials, geometries, and process types.

2.2.2.6 Challenges and Considerations

While RSM is highly powerful, there are challenges:

- Model assumptions: Assumes a reasonably smooth and continuous response surface.
- Data variability: Natural fibre heterogeneity may introduce noise into the system.
- Factor range: Selecting inappropriate factor ranges can lead to extrapolation errors.
- Overfitting: Including non-significant terms may result in models that lack predictive accuracy.

Careful experimental planning, data validation, and iterative refinement of models are essential to overcome these limitations.

RSM offers a scientific, efficient, and statistically robust way to model and optimize the FDM-based green composite fabrication process. By capturing interactions and non-linearities, RSM enhances our understanding of how different process variables influence key product attributes. Its ability to minimize waste, reduce costs, and maximize performance makes it an indispensable tool in the modern toolkit of materials engineers and sustainable manufacturing researchers. As natural fibre-reinforced composites continue to gain prominence, RSM will be at the forefront of innovation, driving the development of high-performance, eco-friendly products in AM.

2.2.3 Analysis of Variance

ANOVA is a fundamental statistical technique used to determine whether there are statistically significant differences between the means of different groups or treatment combinations. In the domain of experimental design, particularly in FDM of green composites, ANOVA plays a crucial role in analyzing how various process parameters affect the output performance metrics (such as tensile strength, surface roughness, or dimensional stability) and in assessing the adequacy of predictive models derived from RSM or regression analysis. Green composites, which combine bio-based matrices (e.g., PLA, PHB) with natural fibres (e.g., jute, hemp, flax), are subject to high variability in both material properties and process outcomes. ANOVA provides a rigorous statistical framework to analyze experimental data, quantify factor effects, identify significant interactions, and validate empirical models, ensuring that observed results are not due to random variation.

2.2.3.1 Purpose of ANOVA in Green Composite Research

In FDM-based green composite modelling, ANOVA serves several key functions:

- Test the significance of each input factor and interaction term.
- Partition total variability in the data into meaningful sources (e.g., factor effects, interactions, error).
- Assess model adequacy in regression or response surface models.
- Control experimental error by determining the reproducibility and variability across trials.
- Rank factors based on their influence on output responses.

By doing so, ANOVA helps researchers and engineers make data-driven decisions regarding process optimization and material design.

2.2.3.2 Basic Concepts and Terminology in ANOVA

Sources of Variation: In a typical DOE setup involving multiple factors (e.g., nozzle temperature, fibre content, print speed), the total variation in the response (e.g., tensile strength) can be decomposed into the following components:

- Sum of Squares for Factors (SSA, SSB, etc.): Variation due to individual factors.
- Sum of Squares for Interaction (SSAB, SSAC, etc.): Variation due to interaction effects between factors.
- Sum of Squares for Error (SSE): Residual variation not explained by the model (experimental noise).
- Total Sum of Squares (SST): Overall variation in the dataset.

$$SST = SSA + SSB + SSAB + \cdots + SSE$$

Degrees of Freedom (DF): DF represent the number of independent pieces of information used to estimate a parameter or variation. Each source of variation has its own DF, and the total DF is the sum of the individual ones.

Mean Square (MS): MS is calculated by dividing each Sum of Squares by its respective DF:

$$MS = \frac{SS}{DF}$$

F-Statistic: The F-value (F-ratio) is used to test whether the means of different groups are significantly different:

$$F = \frac{MS_{Factor}}{MS_{Error}}$$

A high F-value indicates that the factor has a significant effect on the response variable.

p-Value: The p-value is the probability of obtaining an F-value as extreme as, or more extreme than, the observed value, under the null hypothesis (that the factor has no effect). A commonly used threshold is:

- $p < 0.05$: Statistically significant
- $p < 0.01$: Highly significant
- $p > 0.05$: Not significant

2.2.3.3 Application of ANOVA in FDM-Based Green Composite Modelling

Let us consider a typical case: optimizing the tensile strength of PLA/hemp fibre composites fabricated using FDM. Factors include:

- A: Nozzle temperature
- B: Fibre content
- C: Layer height

After conducting experiments using a CCD, a regression model is developed. To evaluate the significance of model terms and overall model performance, ANOVA is performed (Table 2.2).

From Table 2.2, we can interpret the following:

- The model is statistically significant ($p < 0.05$).
- Nozzle temperature and fibre content are highly significant factors.

TABLE 2.2   ANOVA table example

SOURCE	SUM OF SQUARES	DEGREES OF FREEDOM	MEAN SQUARE	F-VALUE	P-VALUE
Model	180.56	6	30.09	12.84	0.0012
A (Temperature)	70.24	1	70.24	29.96	0.0003
B (Fibre content)	45.12	1	45.12	19.24	0.0011
C (Layer height)	12.23	1	12.23	5.23	0.047
AB	15.43	1	15.43	6.66	0.038
AC	10.32	1	10.32	4.23	0.061
Residual error	18.72	8	2.34		
Total	199.28	14			

Abbreviations: ANOVA, analysis of variance.

- Layer height is marginally significant ($p \approx 0.05$).
- The interaction between temperature and fibre content (AB) is significant.

2.2.3.3.1 Model Adequacy Checks
- R^2 (Coefficient of Determination): Indicates the proportion of variability explained by the model.
- Adjusted R^2: Adjusted for the number of predictors.
- Predicted R^2: Indicates how well the model predicts new observations.

2.2.3.4 Types of ANOVA Models

2.2.3.4.1 One-Way ANOVA
Used when there is a single factor with multiple levels. Example: Analyzing the effect of different fibre types (e.g., jute, flax, kenaf) on tensile strength.

2.2.3.4.2 Two-Way ANOVA
Used for experiments with two factors. Example: Evaluating the combined effect of fibre content and raster angle.

2.2.3.4.3 Multifactor ANOVA
Used in full factorial or fractional factorial DOE involving three or more factors. It is the standard approach in RSM-based modelling.

2.2.3.5 ANOVA Assumptions and Validity

To ensure valid inferences from ANOVA, the following assumptions should be satisfied:

- Independence of observations.
- Normality of residuals.
- Homogeneity of variance (equal variance across treatments).

Violations can be checked using:

- Normal probability plots of residuals.
- Levene's test or Bartlett's test for equality of variances.
- Residual vs. fitted value plots to detect non-constant variance.

If assumptions are violated, data transformations (e.g., Box-Cox) or non-parametric methods (e.g., Kruskal–Wallis test) may be used.

2.2.3.6 Visualization of ANOVA Results

- Interaction Plots: Show how the effect of one factor depends on the level of another.
- Main Effects Plots: Show the average response at each level of a factor.
- Pareto Charts: Display the magnitude and significance of factor effects.
- Residual Plots: Used to check ANOVA assumptions and model adequacy.

2.2.3.7 Advantages of Using ANOVA in Green Composite Modelling

- Quantifies significance: Helps identify which parameters most influence the desired output.
- Eliminates guesswork: Enables decisions based on statistical evidence, not intuition.
- Supports optimization: Forms the basis for model selection and process refinement in RSM.
- Assesses reproducibility: Identifies experimental noise and helps improve repeatability.
- Enhances scientific rigor: Essential for publishing high-impact experimental research.

2.2.3.8 Limitations and Considerations

While ANOVA is a powerful technique, it has some limitations:

- Assumes additivity and linearity: Not ideal for highly non-linear systems unless extended to second-order models.
- Sensitive to outliers: Outliers can distort the *F*-value and mislead interpretations.
- Requires balanced design: Unequal sample sizes reduce statistical power.
- Cannot provide detailed predictive capability alone: Must be combined with regression or RSM for modelling.

2.2.4 Taguchi Methods and Robust Parameter Design

In manufacturing systems, product performance is not only influenced by controllable design and process parameters but also by uncontrollable external factors or "noise" such as environmental conditions, material inconsistencies, or operator variability. To ensure high-quality performance that remains stable under varying real-world conditions, Genichi Taguchi developed a powerful methodology focused on robust design, the ability to make products and processes insensitive to noise. The Taguchi method combines the principles of DOE with robust parameter optimization strategies. In the context of FDM of green composites, which involve bio-based, often heterogeneous materials like jute, flax, kenaf, or hemp fibres, Taguchi methods provide a structured yet economical approach to identify optimal process settings that deliver consistent product quality while minimizing material waste and environmental footprint.

2.2.4.1 Fundamentals of the Taguchi Method

The Taguchi approach focuses on:

- Quality through robust design rather than inspection.
- Minimizing variation due to uncontrollable noise.
- Using orthogonal arrays (OAs) for experimental planning to study the influence of multiple factors efficiently.

Key goals of Taguchi's methodology are:

- To identify control factors (e.g., nozzle temperature, fibre content, layer height) that influence product quality.
- To make designs less sensitive to noise, thereby enhancing reliability and sustainability.
- To optimize process parameters by maximizing the signal-to-noise (S/N) ratio, a measure of robustness.

2.2.4.2 Components of Taguchi Methods

2.2.4.2.1 Control Factors
These are the variables that can be set and controlled in the process (e.g., printing speed, layer height, raster angle).

2.2.4.2.2 Noise Factors
These are uncontrollable in real-world scenarios (e.g., humidity, ambient temperature, raw material inconsistency). In robust design, the aim is to minimize the effect of these noise factors.

2.2.4.2.3 Orthogonal Arrays
Taguchi uses OAs to design experiments in a way that balances the influence of factors across trials while reducing the number of experiments. Each column in an OA represents a factor, and each row represents a unique combination of factor levels.
Example:

- L_9 OA: Three factors at three levels → nine trials
- L_{18} OA: Can accommodate a mix of two- and three-level factors.

S/N Ratio: The S/N ratio quantifies the desirability of an output by considering both the mean and the variability. Depending on the goal, one of the following formulations is used:

- Larger-the-better (e.g., strength):

$$\frac{S}{N} = -10\log_{10}\left(\frac{1}{n}\sum_{i=1}^{n}\frac{1}{y_i^2}\right)$$

- Smaller-the-better (e.g., surface roughness):

$$\frac{S}{N} = -10\log_{10}\left(\frac{1}{n}\sum_{i=1}^{n}y_i^2\right)$$

- Nominal-the-best (e.g., dimension accuracy):

$$\frac{S}{N} = -10\log_{10}\left(\frac{1}{n}\sum_{i=1}^{n}\frac{\bar{y}^2}{s^2}\right)$$

where:

- y_i: observed response values,
- $\bar{y}$: mean of observed values,
- s: standard deviation,
- n: number of observations.

2.2.4.3 Application of Taguchi Methods in FDM-Based Green Composite Modelling

The FDM process, especially when using green composites, is prone to variability due to:

- Natural fibre inhomogeneity
- Moisture sensitivity
- Thermal degradation risk
- Irregular dispersion of fibres

These factors make robust optimization critical. Taguchi methods are used to:

2.2.4.3.1 Identify Key Process Parameters
For example, selecting the most critical parameters that influence interlayer bonding strength in a PLA-jute composite, such as:

- Nozzle temperature
- Bed temperature
- fibre weight fraction
- Print speed

2.2.4.3.2 Optimize Performance under Noisy Conditions
By introducing simulated noise (e.g., humidity variation, fluctuating fan speeds), the robustness of different parameter combinations can be tested.

2.2.4.3.3 Reduce Experimental Burden
A full factorial design for four factors at three levels would require 81 runs (3^4), but an L_9 or L_{27} OA can study the same parameters using only 9 or 27 trials, respectively, with reasonable resolution.

2.2.4.4 Taguchi Analysis Process: Step-by-Step

2.2.4.4.1 Step 1: Selection of Factors and Levels

Choose control factors (e.g., nozzle temperature, fibre content) and assign two to three levels per factor based on prior knowledge or literature.

2.2.4.4.2 Step 2: Selection of an Appropriate OA

Based on the number of factors and levels, choose a suitable OA (e.g., L_9, L_{18}, L_{27}).

2.2.4.4.3 Step 3: Conducting the Experiments

Perform the experiments as per the OA matrix, recording the relevant responses (e.g., tensile strength, flexural strength, surface roughness).

2.2.4.4.4 Step 4: Calculating S/N Ratios

Compute the *S/N* ratio for each response based on the desired performance (larger, smaller, or nominal).

2.2.4.4.5 Step 5: Response Analysis

- Plot main effects to identify optimal factor levels.
- Rank factors by their delta (difference between highest and lowest *S/N* ratio values).
- Use ANOVA to determine which factors are statistically significant.

2.2.4.4.6 Step 6: Confirmation Experiment

Verify the optimal settings predicted by the analysis through additional experimental runs.

2.2.4.5 Case Example: Taguchi-Based Optimization in PLA-Flax FDM Green Composites

In an experiment to improve the tensile strength of a PLA-flax FDM composite, the following setup was used:

- Factors and Levels:
  - A: Nozzle temperature (190°C, 200°C, 210°C)
  - B: Fibre content (5%, 10%, 15%)
  - C: Infill density (40%, 60%, 80%)
- OA used: L_9 (3^4) OA
- Response: Tensile strength (MPa)

- Results:
  - The optimal combination was A_2 (200°C), B_1 (5%), C_3 (80%)
  - S/N analysis showed nozzle temperature had the greatest influence, followed by infill density.
  - Confirmation test showed a 12% improvement in strength with reduced variability across trials.

2.2.4.6 Advantages of Taguchi Methods in Green Composite Modelling

- Cost-Effective: Reduces the number of experimental runs.
- Noise-Resistant: Designs robust products that perform well under variation.
- Structured and Systematic: Provides clear guidelines for experiment planning.
- Scalable: Can be applied from lab-scale trials to industrial production.
- Intuitive Optimization: Easy-to-interpret S/N ratio plots for decision-making.

2.2.4.7 Limitations and Considerations

- Interaction Effects Not Always Captured: Taguchi's approach assumes minimal interaction, which may be limiting in highly interactive systems like FDM.
- Factor Level Discreteness: Works best with two to three levels; modelling curvature (non-linear responses) requires extensions like RSM.
- Fixed OA Structure: Factor levels and OA configurations are rigid; not flexible for every problem.

To overcome these issues, hybrid approaches (Taguchi + RSM or Taguchi + Grey Relational Analysis [GRA]) are sometimes used.

2.2.4.8 Extensions and Advanced Applications

- Robust Design with Multiple Responses: Using multi-response optimization techniques like GRA integrated with Taguchi to optimize conflicting criteria (e.g., strength ↑, roughness ↓).
- Software Support: Tools like Minitab, Design-Expert, and JMP offer built-in templates for Taguchi analysis.

The Taguchi method simplifies optimization by using OAs for designing experiments. It focuses on enhancing product or process robustness by minimizing variability due to uncontrollable factors (noise).

In FDM green composite modelling:

- Taguchi's *S/N* ratios are used to measure robustness.
- The method has been used to optimize process parameters like print temperature, infill pattern, and feed rate in PLA/flax composites under varying humidity conditions.

Advantage: Taguchi is less computationally intensive than full factorial DOE and is particularly useful in industrial settings with limited experimentation budgets.

2.2.5 Regression Analysis

Regression analysis is a powerful statistical tool used to examine the relationship between one or more independent variables (input parameters) and a dependent variable (response). It is widely applied in green composite research to develop empirical models that describe how changes in process parameters influence the performance characteristics of natural fibre-reinforced thermoplastics, especially in AM processes like FDM. In FDM-based green composite fabrication, where materials such as PLA, PHB, or polycaprolactone are reinforced with fibres like jute, hemp, flax, or bamboo, multiple factors interact in complex ways. Regression analysis helps researchers uncover underlying patterns, quantify factor contributions, and develop predictive equations for optimizing properties such as modulus, impact strength, dimensional accuracy, and degradability.

2.2.5.1 Purpose of Regression Analysis in Green Composite Research

The primary goals of using regression analysis in this context are to:

- Model the relationship between process inputs (e.g., print temperature, layer height, fibre content) and output responses (e.g., tensile strength, flexural modulus, surface quality).
- Predict performance under new or untested conditions.
- Identify key factors affecting responses, facilitating informed decision-making.

- Support optimization through mathematical modelling and simulation.
- Assess the relative importance of each input variable and their interactions.

These empirical models enable scientists and engineers to design experiments more efficiently, improve repeatability, and accelerate process development.

2.2.5.2 Types of Regression Techniques Used

2.2.5.2.1 Linear Regression

Linear regression is the simplest form of regression and assumes a direct proportional relationship between the input variable(s) and the response.

$$Y = \beta_0 + \beta_1 X + \varepsilon Y$$

where:

- Y = dependent variable (e.g., tensile strength),
- X = independent variable (e.g., fibre content),
- β_0 = intercept,
- β_1 = slope (rate of change of Y with X),
- ε = random error.

2.2.5.2.2 Application

Linear regression is useful in early studies where the relationship between a single parameter and the response is expected to be monotonic and straightforward. For example, evaluating how increasing fibre weight fraction affects the modulus of elasticity of PLA-flax composites.

2.2.5.3 Multiple Linear Regression

Multiple linear regression (MLR) extends linear regression to multiple predictors. It models the combined linear influence of two or more independent variables on a single response.

$$Y = \beta_0 + \beta_1 X_1 + \beta_2 X_2 + \cdots + \beta_k X_k + \varepsilon_Y$$

where:

- $X_1, X_2, \ldots, X_k$; k = input variables (e.g., print temperature, raster angle, fibre size),
- $\beta_1, \ldots, \beta_k$ = regression coefficients.

2.2.5.3.1 Application

MLR is widely used in modelling FDM-based green composites where multiple factors affect the response. For instance, a model predicting impact strength of hemp-reinforced PLA can incorporate both raster angle and fibre content as predictors.

2.2.5.3.2 Benefits

- Accounts for simultaneous factor effects.
- Enables main effects screening.
- Enhances predictive accuracy for process planning.

2.2.5.3.3 Software Tools

Packages such as Minitab, Design-Expert, SPSS, R, and Python (scikit-learn, stats models) are commonly used to perform MLR analysis.

2.2.5.4 Polynomial Regression

Polynomial regression is used when the response surface exhibits non-linear behaviour. It incorporates higher-degree terms (squared, cubic, etc.) of the input variables.

$$Y = \beta_0 + \beta_1 X + \beta_2 X_2 + \beta_3 X_3 + \cdots + \varepsilon Y$$

In multivariate form (quadratic for two variables):

$$Y = \beta_0 + \beta_1 X_1 + \beta_2 X_2 + \beta_{11} X_{12} + \beta_{22} X_{22} + \beta_{12} X_1 X_2 + \varepsilon Y$$

2.2.5.4.1 Application

Polynomial regression is often used within RSM for:

- Modelling non-linear relationships.
- Capturing interactions between variables.
- Locating optimal process settings.

In green composite FDM, it helps in understanding how combined changes in fibre content, temperature, and layer height affect dimensional stability or strength-to-weight ratio.

2.2.5.4.2 Example

A second-order polynomial model for the flexural strength of a PLA-jute composite as a function of nozzle temperature (X_1) and infill density (X_2) might look like:

$$\text{Flexural Strength} = 25 + 2.1 X_1 - 1.5 X_2 + 0.03 X_{12} + 0.04 X_{22} - 0.07 X_1 X_2$$

2.2.5.5 Model Development and Evaluation Process

2.2.5.5.1 Step 1: Data Collection
Data from DOE or empirical trials is gathered by varying key parameters across a pre-defined range.

2.2.5.5.2 Step 2: Model Fitting
Regression models are fitted using the least squares method, minimizing the sum of squared differences between predicted and observed responses.

2.2.5.5.3 Step 3: Model Validation
Model accuracy is validated using:

- Coefficient of Determination (R^2): Measures how well the model explains the variance in data. Values >0.8 generally indicate a good fit.
- Adjusted R^2: Adjusts R^2 based on the number of predictors; prevents overfitting.
- p-Values: Indicate statistical significance of coefficients (typically, $p < 0.05$ is significant).
- Root Mean Square Error (RMSE): Indicates prediction error.
- Residual Plots: Check for homoscedasticity and normality.

2.2.5.5.4 Step 4: Prediction and Optimization
Once validated, the model can be used to:

- Predict responses for new combinations of input variables.
- Perform what-if analyses.
- Support multi-objective optimization using desirability functions or evolutionary algorithms.

2.2.5.6 Applications in FDM-Based Green Composites

2.2.5.6.1 Predicting Mechanical Properties
- Tensile and flexural strength prediction using MLR and polynomial regression based on:
  - Fibre type and loading
  - Print speed
  - Bed/nozzle temperature

2.2.5.6.2 Surface Quality Modelling
- Use regression to relate surface roughness or dimensional error to print layer height and infill density.

- Modelling biodegradation rates, carbon footprint, or material utilization efficiency using regression to optimize eco-performance.

2.2.5.7 Example Case Study

Study Objective: To model the impact strength of a hemp-reinforced PLA green composite.

Factors considered:

- Raster angle (°)
- Fibre weight percentage (%)
- Nozzle temperature (°C)

Regression Technique: MLR
    Derived Model:

$$\text{Impact Strength} = 8.5 + 0.2\left(\text{Raster Angle}\right) + 1.1\left(\text{Fiber Content}\right) \\ - 0.05\left(\text{Nozzle Temp}\right)$$

Interpretation:

- Impact strength increases with raster angle and fibre content.
- Higher nozzle temperatures beyond a threshold may degrade the natural fibre, reducing impact strength.

Model Evaluation:

- $R^2 = 0.91$
- Adjusted $R^2 = 0.88$
- RMSE = 0.32

2.2.5.8 Challenges and Considerations

- Overfitting: Including too many terms or irrelevant predictors can make the model complex and inaccurate for new data.
- Multicollinearity: When predictors are highly correlated, it can distort the regression coefficients.
- Non-Linearity: Linear models may fail to capture curved trends; polynomial terms or transformation may be needed.

- Material Variability: In green composites, inherent variability in natural fibres can introduce unpredictability in modelling.

2.2.6 Principal Component Analysis

Principal component analysis (PCA) is a widely used multivariate data analysis technique that simplifies complex datasets by reducing their dimensionality while retaining most of the original variability. It does so by transforming a set of correlated variables into a new set of uncorrelated variables called principal components (PCs). These PCs are linear combinations of the original variables and are ordered in such a way that the first few PCs capture most of the variance in the data. In the field of green composite modelling, particularly in FDM processes using natural fibre-reinforced thermoplastics, PCA plays a crucial role in managing and interpreting the multivariate nature of the data. These datasets often consist of mechanical, thermal, morphological, rheological, and processing parameters, each of which can be influenced by a variety of material and process factors. PCA helps simplify, interpret, and visualize these complex relationships.

2.2.6.1 Why PCA in Green Composite Research?

The fabrication of green composites involves many variables, including:

- Input parameters: fibre content, fibre length, print speed, nozzle temperature, raster angle, build orientation.
- Material properties: modulus, tensile/flexural strength, elongation at break, thermal conductivity, degradation temperature.
- Morphological features: porosity, fibre dispersion, interfacial adhesion.
- Environmental metrics: biodegradability, recyclability, CO_2 footprint.

PCA is especially useful when:

- The dataset has many correlated variables.
- The aim is to explore underlying patterns or relationships.
- The researcher wants to visualize class/group differences (e.g., based on fibre types or treatments).
- There's a need to reduce the dimensionality before applying further modelling techniques (e.g., clustering, regression, or classification).

2.2.6.2 The Mathematics behind PCA

PCA involves the following steps:

2.2.6.2.1 Standardization

Variables are often measured in different units (e.g., MPa, °C, %, mm), so data is standardized (zero mean, unit variance) to ensure comparability.

2.2.6.2.2 Covariance Matrix Calculation

The covariance matrix shows how variables relate to each other (i.e., their degree of correlation).

Eigen Decomposition: The eigenvalues and eigenvectors of the covariance matrix are computed.

- Eigenvalues indicate the amount of variance explained by each PC.
- Eigenvectors (principal axes) define the direction of maximum variance.

Transformation: Original data is projected onto the eigenvectors to generate the PC scores.

2.2.6.3 Applications of PCA in Green Composite Modelling

2.2.6.3.1 Feature Extraction and Data Compression

PCA reduces a complex dataset (e.g., 10+ input features) to a smaller set of uncorrelated PCs, preserving key information.

Example: Instead of analyzing tensile strength, flexural strength, and elongation separately, PCA may reveal that PC1 represents overall mechanical integrity while PC2 captures ductility-specific behaviour.

This makes the data easier to:

- Model
- Visualize
- Interpret

2.2.6.3.2 Identifying Hidden Patterns and Correlations

PCA can reveal latent variables or underlying patterns in the data that are not obvious from univariate or bivariate analysis.

- Example: PCA may show that thermal degradation temperature and fibre-matrix interfacial strength co-vary in such a way that they form a PC representing thermal-mechanical robustness.

Such insights can help:

- Guide material formulation strategies.
- Select optimal fibre types or treatments.
- Understand failure mechanisms.

2.2.6.3.3 Visualization of Multivariate Data
PCA allows for 2D or 3D scatter plots of PCs, helping to visualize:

- Clustering of materials by fibre type (e.g., jute vs. hemp)
- Processing conditions (e.g., low vs. high nozzle temperature)
- Outliers or anomalies in the dataset

This is valuable in quality control, exploratory data analysis, and pattern recognition.

2.2.6.3.4 Pre-Processing for Other Modelling Techniques
PCA is often used before:

- Clustering algorithms (e.g., k-means, hierarchical clustering)
- Regression models (e.g., PC regression)
- Classification algorithms (e.g., SVM, decision trees)

By reducing noise and multicollinearity, PCA improves the robustness and accuracy of these models.

2.2.6.4 Example: PCA in Natural Fibre-Reinforced PLA Composites

2.2.6.4.1 Dataset
A research team collected the following parameters for 30 samples of FDM-printed green composites made from PLA reinforced with jute, hemp, or flax:

- Tensile strength
- Flexural strength

- Modulus
- Elongation at break
- Thermogravimetric analysis (TGA) degradation temperature
- Surface roughness
- Fibre pull-out length (from scanning electron microscopy [SEM] images)
- Moisture absorption rate

2.2.6.4.2 PCA Findings
- PC1 (explaining 45% of variance) captured mechanical strength features (tensile, flexural, modulus).
- PC2 (explaining 25%) represented thermal and moisture stability (TGA temp, water uptake).
- Samples grouped into three clusters corresponding to fibre type, with jute-based composites showing higher PC1 scores but lower PC2 (i.e., good strength but poor thermal stability).

The PCA biplot helped visualize:

- Correlations (e.g., modulus and flexural strength were strongly aligned)
- Trade-offs (e.g., materials with higher elongation at break had lower modulus)
- Material selection strategies based on application requirements.

2.2.6.5 Challenges and Considerations

- Interpretation Complexity: PCs are mathematical constructs and may not always have direct physical meaning.
- Loss of Information: While PCA retains most of the variance, some subtle but important information may be lost if too few components are retained.
- Linear Assumption: PCA assumes linear relationships among variables; it may not capture complex non-linear dependencies.
- Sensitivity to Scaling: Results can vary significantly depending on whether data is standardized or normalized.

2.2.6.6 PCA Tools and Software for Implementation

- Python: scikit-learn (PCA module), pandas, matplotlib for visualization
- R: prcomp(), ggbiplot, factoextra
- MATLAB: pca(), biplot()

- Minitab, JMP, SPSS: GUI-based PCA modules
- Design-Expert: Often used in conjunction with RSM and DOE

2.3 CONSTRUCTION AND VALIDATION OF PREDICTIVE MODELS FOR MECHANICAL AND FUNCTIONAL PROPERTIES

In the context of FDM for green composites, the fabrication process is influenced by multiple interdependent parameters [15]. These include printing temperature, feed rate, raster angle, fibre content, and fibre-matrix adhesion characteristics. The resultant mechanical and functional properties, such as tensile strength, impact resistance, thermal stability, and biodegradability, are governed by these process–material interactions. To reduce dependency on costly, time-consuming, and resource-intensive experimental trials, researchers and engineers increasingly rely on predictive models that can estimate material behaviour based on input settings. These models serve multiple roles: they assist in understanding underlying phenomena, optimizing process parameters, and ensuring consistent performance for eco-designed products.

Constructing and validating such predictive models requires a structured approach that integrates experimental design, statistical or ML methods, model adequacy checks, and validation through real or simulated trials. The goal is to ensure that the model not only fits the experimental data well but also generalizes effectively to unseen combinations of input variables, thereby supporting the design for sustainability, reliability, and efficiency.

2.3.1 Model Construction: General Framework

The process of constructing a predictive model involves several stages, each integral to developing a robust and accurate representation of the system being studied.

2.3.1.1 Defining the Objective

The first step is to clearly define what property is to be predicted (e.g., tensile strength, flexural modulus, thermal conductivity, surface roughness, or biodegradation rate). The mechanical properties typically represent structural behaviour, while functional properties may refer to performance under operational conditions (e.g., thermal resilience or eco-toxicity).

2.3.1.2 Selection of Input Variables (Features)

Variables are selected based on domain knowledge, literature evidence, and preliminary screening experiments. In green composite FDM modelling, typical input parameters may include:

- Fibre volume fraction
- Printing temperature
- Bed temperature
- Raster angle
- Layer height
- Infill density
- Print speed
- Material blend ratio (e.g., PLA + lignin or PLA + flax)

Careful pre-processing of the data is required to eliminate redundant, irrelevant, or collinear variables that may affect model performance.

2.3.1.3 Experimental Design and Data Collection

Data for model construction may be gathered through:

- DOE frameworks (e.g., full factorial, Taguchi, RSM).
- Simulated datasets from finite element modelling or computational fluid dynamics.
- Sensor-based process monitoring (e.g., thermographic, or acoustic emission data).

High-quality datasets with adequate sampling, replication, and randomization improve model robustness.

2.3.1.4 Selection of Modelling Approach

The choice of modelling technique depends on the nature of the dataset (size, complexity, noise) and the prediction goal (interpretability vs. accuracy). Common techniques include:

- Linear and Polynomial Regression: For systems with simple or second-order interactions.

- ANNs: For complex, non-linear datasets.
- SVMs: For high-dimensional pattern recognition.
- Random Forest and Decision Trees: For classification or regression tasks with interpretability.
- Gaussian Process Models: For probabilistic prediction with uncertainty estimation.
- Principal Component Regression or Partial Least Squares Regression: When multicollinearity is an issue.

2.3.2 Model Construction for Mechanical Properties

Predicting Tensile and Flexural Strength: Tensile and flexural properties are critical for load-bearing applications of green composites. Models are built using input parameters like:

- Fibre content (wt%)
- Nozzle temperature (°C)
- Raster angle (°)
- Layer height (mm)

For example, a second-order polynomial model may be developed through RSM:

$$\text{Tensile Strength} = \beta_0 + \sum_{i=1}^{n} \beta_i X_i + \sum_{i=1}^{n} \beta_{ii} X_i^2 + \sum_{i=1}^{n} \beta_{ij} X_i X_j + \varepsilon$$

where X_i are the process variables, and ε represents the residual error.

Modelling Interlayer Adhesion and Toughness: Green composites often suffer from poor interlayer adhesion due to moisture sensitivity or fibre misalignment. To predict interlayer bonding strength:

- ANNs can be trained using microstructural inputs (e.g., porosity levels from SEM images).
- SVM models can classify fracture modes (ductile vs. brittle) based on printing conditions.

Such predictive insights can guide process modification to enhance part reliability.

2.3.3 Model Construction for Functional Properties

Thermal and Biodegradation Modelling: Functional properties such as thermal stability, ignition resistance, or biodegradability are especially relevant for eco-composite applications.

- TGA data can be modelled using regression or multivariate calibration techniques to link thermal degradation temperatures to fibre/matrix ratios and filler types.
- Biodegradation rate models can relate the decay curve to composite composition and environmental exposure conditions using exponential decay functions or Weibull distribution fits.

Moisture Absorption and Aging: Moisture uptake affects dimensional accuracy and strength in natural-fibre composites. Predictive models using Fickian diffusion equations combined with ML approaches can help forecast long-term performance.

2.3.4 Model Validation

Once a predictive model is built, it must be rigorously validated to ensure accuracy, reliability, and generalizability. Validation involves both **internal checks** using training data and external validation using test data or real-world experiments.

2.3.4.1 Cross-Validation Techniques

- k-Fold Cross-Validation: Data is split into k subsets; each subset is used once as a test set while the remaining are used for training. Results are averaged for performance metrics.
- Leave-One-Out Cross-Validation: Each data point is used as a test case once; suitable for small datasets.
- Train/Test Split: A common approach where 70–80% of the data is used for training and the rest for testing.

2.3.4.2 Performance Metrics

Performance is quantified using:

- R^2 (Coefficient of Determination): Indicates the proportion of variance explained.

- RMSE: Measures average error magnitude.
- Mean Absolute Error: Provides an error measure that is less sensitive to outliers.
- Mean Absolute Percentage Error: Useful for relative error comparison.
- Residual Analysis: Checks for model bias, variance consistency, and normality.

2.3.4.3 Experimental Validation

After computational validation, confirmation runs are conducted to compare predicted vs. actual outcomes under new process settings. This ensures that the model has real-world relevance.

2.3.4.4 Model Robustness and Generalization

Robust models should:

- Predict well on new data.
- Be insensitive to noise.
- Show low overfitting (high training accuracy but poor test performance is a red flag).

Advanced validation methods, such as bootstrapping or Monte Carlo simulations, can be employed to assess the model's generalization ability under uncertainty.

2.3.5 Integration with Optimization

Predictive models often serve as a foundation for optimization algorithms that search for the best combination of inputs to maximize or minimize desired outputs.

- Desirability-Based Multi-Objective Optimization: Combines predictive models with desirability functions for trade-off decisions.
- Genetic algorithm (GA), particle swarm optimization, and simulated annealing (SA) can be linked to regression or ANN models for global optimization.

Example:
A tensile strength model is combined with a biodegradability model to find the optimal print parameters that balance structural strength with environmental degradation goals.

2.3.6 Case Study: Predicting Impact Strength of PLA-Jute Composite

A team conducted 27 experiments using BBD to vary:

- Fibre content (10–30%)
- Nozzle temperature (180–220°C)
- Raster angle (0°, 45°, 90°)

Using RSM, the model obtained:

$$\text{Impact Strength} = 12.3 + 1.7A + 0.9B - 0.5C - 0.2AB + 0.1AC + 0.3BC - 0.4A^2 - 0.3B^2 - 0.1C^2$$

Validation:

- $R^2 = 0.92$; Adjusted $R^2 = 0.89$
- RMSE = 0.34 J/m
- Confirmation tests showed deviation <5%

Outcome: The model enabled the researchers to recommend optimal fibre content (22%) and printing temperature (205°C) for applications requiring both toughness and environmental performance.

2.3.7 Challenges in Model Construction and Validation

- Material Variability: Natural fibres have inconsistent properties (e.g., moisture content, diameter), introducing uncertainty.
- Data Scarcity: Limited data points can constrain model accuracy, especially for ANN and ML models.
- Multicollinearity: Correlated inputs (e.g., infill density) can distort regression models.
- Overfitting Risks: Especially with high-dimensional or non-linear models.
- Dynamic Behaviours: Real-time process variables may not be fully captured by static models.

Mitigation strategies include:

- Collecting more data under varied conditions
- Using regularization techniques (e.g., Lasso, Ridge)
- Feature selection via PCA or correlation filtering

The construction and validation of predictive models form the cornerstone of data-driven design and optimization in green composite manufacturing via FDM. From simple regression techniques to complex ML algorithms, these models empower researchers to forecast performance, identify key process levers, and balance competing requirements such as strength, cost, and sustainability. A robust modelling framework must include careful input selection, quality data collection, the use of appropriate modelling algorithms, and thorough validation procedures. Only then can predictive models move from theoretical tools to practical instruments of innovation, enabling the scalable and sustainable use of bio-based materials in advanced manufacturing.

2.4 INTEGRATION OF STATISTICAL MODELS INTO SUSTAINABLE DESIGN FRAMEWORKS AND DSSS

As global emphasis intensifies on the transition to a low-carbon and circular economy, sustainable design frameworks are increasingly being adopted in the engineering, manufacturing, and materials domains. These frameworks aim to incorporate environmental, social, and economic considerations at every stage of the design process, from material selection and processing to product use and end-of-life. In this context, green composites, typically consisting of biodegradable or recyclable polymers reinforced with natural fibres, play a pivotal role in enabling sustainable manufacturing. However, sustainable design is not purely conceptual; it must be supported by quantitative decision-making tools that allow engineers to evaluate, compare, and optimize design alternatives systematically [16,17]. This is where statistical models become invaluable. When embedded within DSSs, statistical models serve as predictive engines, performance assessors, and optimization tools. They bridge the gap between raw data, stakeholder goals, and actionable insights, thus operationalizing sustainability principles in real-world manufacturing contexts. This section explores how statistical modelling, including regression, ANOVA, DOE, multivariate analysis, and ML, can be integrated with sustainable design frameworks and DSSs, with a special emphasis on green composite materials and FDM processes.

2.4.1 Sustainable Design Frameworks: A Brief Overview

Sustainable design frameworks are structured methodologies that guide the creation of products with minimal environmental impact and maximum social and economic benefit. Key principles include:

- Life Cycle Thinking: Evaluating the impacts of a product from cradle to grave.
- Eco-Design: Embedding environmental considerations into engineering design.
- Design for Sustainability: Incorporating energy efficiency, recyclability, renewable materials, and social responsibility.
- Design for Additive Manufacturing: Leveraging 3D printing to minimize waste, customize products, and decentralize production.

These frameworks require quantitative metrics, such as embodied energy, carbon footprint, tensile strength per unit energy, and recyclability index—to make data-driven design decisions. Statistical models are crucial in generating, analyzing, and interpreting these metrics.

2.4.2 Role of Statistical Models in Sustainable Product Design

Statistical models can be embedded in sustainable design frameworks to serve several functions.

2.4.2.1 Predictive Modelling

Statistical models help predict product properties (e.g., mechanical strength, dimensional accuracy, biodegradability) based on design and process parameters. These models reduce the need for exhaustive experimentation. For example, a multiple regression model can predict the tensile strength of a PLA-hemp composite based on nozzle temperature, layer height, and fibre weight fraction. This model can then be used to select processing parameters that maximize strength while minimizing energy use and material cost.

2.4.2.2 Optimization

Using tools like RSM or Taguchi methods, statistical models identify parameter combinations that yield optimal responses across multiple objectives, such as maximizing strength and biodegradability while minimizing energy consumption.

2.4.2.3 Trade-Off Analysis

In sustainable design, trade-offs are common: increasing mechanical strength may reduce biodegradability; using high infill densities may enhance performance but increase material consumption. Pareto frontiers and multi-objective optimization models help navigate such conflicts by providing sets of non-dominated solutions.

2.4.2.4 Sensitivity and Uncertainty Analysis

Statistical models allow designers to assess how sensitive product performance is to variations in input parameters—helping to design more robust and reliable systems under uncertainty.

2.4.3 Integration into DSSs

A DSS is a proof-of-concept software-based platform that can aid users in making informed and data-driven choices. Integration of statistical models into DSS involves several components.

2.4.3.1 Data Acquisition Layer

This layer gathers data from:

- Experimental trials
- Real-time sensors (temperature, force, dimensional scanning)
- Historical databases
- Sustainability metrics (LCA data, carbon accounting)

2.4.3.2 Statistical Analysis Module

This module includes:

- Regression models (linear, polynomial, non-linear)
- DOE frameworks
- PCA for dimensionality reduction
- ANOVA for variance partitioning
- ML models (e.g., Random Forest, ANN) for classification and prediction

These tools transform raw data into interpretable patterns, predictive equations, and parameter insights.

2.4.3.3 Decision Logic and Optimization Module

This component integrates the output from statistical models with:

- Multi-criteria decision-making methods such as Analytic Hierarchy Process, Technique for Order Preference by Similarity to Ideal Solution (TOPSIS), or TODIM.
- Multi-objective optimization algorithms (e.g., NSGA-II).
- Desirability functions to rank design alternatives.

2.4.3.4 User Interface and Visualization

Visual dashboards, biplots (from PCA), Pareto charts, and surface plots allow designers to interact with the DSS and make informed decisions quickly.

2.4.4 Case Study: A DSS for Green Composite FDM Part Selection

Imagine a DSS designed to support the selection and processing of green composites for automotive interior panels using FDM.

2.4.4.1 Inputs

- Material options: PLA + flax, PLA + jute, PHB + hemp
- Process variables: print temperature, infill density, raster angle
- Sustainability metrics: embodied energy, volatile organic compound emissions, recyclability index

2.4.4.2 Statistical Model Integration

- DOE is used to conduct structured experiments across fibre-matrix combinations.
- Regression analysis predicts tensile strength and impact resistance from input parameters.
- ANOVA identifies statistically significant factors.
- RSM identifies optimal processing settings.
- PCA reduces dimensionality of multivariate property data.
- TOPSIS ranks material-process combinations based on criteria like strength-to-weight ratio, cost, and environmental score.

2.4.4.3 Output

- Recommended material: PHB + hemp at 210°C with 60% infill and 45° raster angle
- Expected tensile strength: 32 MPa; biodegradability rating: 9/10
- Visual plots: trade-off curve between strength and environmental score

This integration of statistical models within a sustainable design DSS accelerates innovation, improves reproducibility, and reduces ecological impact.

2.4.5 Benefits of Integration

2.4.5.1 Holistic Decision-Making

Combining technical, environmental, and economic parameters allows stakeholders to make decisions aligned with the triple bottom line.

2.4.5.2 Enhanced Efficiency

Reduces trial-and-error, accelerates prototyping, and cuts down waste and energy consumption.

2.4.5.3 Customization

Statistical models can be tailored to different use cases, such as automotive parts, consumer electronics, or biomedical devices made from green composites.

2.4.5.4 Transparency and Reproducibility

Data-driven modelling provides transparency in decision-making and ensures repeatability in product development.

2.4.5.5 Scalability

Once trained and validated, models can be scaled across product lines or geographic markets with minor tuning.

2.4.6 Challenges in Model Integration

2.4.6.1 Data Quality and Availability

Poor or inconsistent data (e.g., from inconsistent natural fibre quality) can reduce model accuracy.

2.4.6.2 Model Overfitting

Complex statistical models may overfit training data and perform poorly in practice.

2.4.6.3 Interdisciplinary Barriers

Bridging the gap between material science, statistics, and software engineering can be challenging.

2.4.6.4 Computational Complexity

Advanced models like neural networks or GA require significant computing resources.

2.4.6.5 Stakeholder Acceptance

Adoption may be hindered by a lack of trust in "black-box" models unless proper explainability tools are integrated.

The integration of statistical models into sustainable design frameworks and DSSs represents a significant step towards data-driven sustainability in engineering design and manufacturing. In the realm of green composites, especially those fabricated using FDM, statistical models enable accurate property prediction, robust optimization, and intelligent trade-off analysis, making it feasible to balance performance, cost, and environmental responsibility. By embedding such models into interactive and dynamic DSS platforms, engineers and designers can make informed, transparent, and agile decisions. As computational capabilities and modelling techniques advance, the seamless fusion of statistical rigor with sustainability imperatives will be key to fostering a resilient, eco-efficient, and innovation-driven industrial landscape.

2.5 CONCLUSION

The transition to sustainable materials and manufacturing practices has positioned green composites at the forefront of eco-conscious innovation. While these materials offer compelling advantages such as biodegradability, renewable sourcing, and reduced environmental burden, they also present complexities arising from their natural variability and processing sensitivity. This chapter has demonstrated that statistical modelling serves as a cornerstone for navigating these challenges, enabling precise characterization, efficient optimization, and predictive insights into green composite behaviour during FDM processing.

By systematically applying tools such as DOE, RSM, ANOVA, Taguchi methods, regression techniques, and PCA, researchers can not only reduce experimental costs but also deepen their understanding of the intricate relationships between process parameters and performance metrics. The construction and validation of empirical and ML models further empower engineers to predict key outcomes—such as mechanical strength, surface finish, and biodegradability—under varied conditions, ensuring product reliability and consistency.

Importantly, this chapter emphasized the integration of statistical models within sustainable design frameworks and DSSs. This synthesis bridges the gap between raw experimental data and high-level design decisions, enabling multi-objective optimization that balances environmental impact, mechanical functionality, and economic viability. As sustainable manufacturing continues to evolve, the convergence of data-driven modelling, real-time process control, and eco-design principles will be essential for fostering intelligent, agile, and responsible innovation in the domain of green composites.

In sum, harnessing the power of statistics is not merely a methodological choice but a strategic necessity in the pursuit of sustainable material development and AM excellence.

REFERENCES

1. Oh, Eunyoung, Marcela María Godoy Zúñiga, Tan Binh Nguyen, Baek-Hwan Kim, Tran Trung Tien, and Jonghwan Suhr. "Sustainable green composite materials in the next-generation mobility industry: Review and prospective." *Advanced Composite Materials* 33, no. 6 (2024): 1368–1419.
2. Fragassa, Cristiano, Felipe Vannucchi de Camargo, and Carlo Santulli. "Sustainable biocomposites: Harnessing the potential of waste seed-based fillers in eco-friendly materials." *Sustainability* 16, no. 4 (2024): 1526.
3. Alarifi, Ibrahim M. "Revolutionising fabrication advances and applications of 3D printing with composite materials: A review." *Virtual and Physical Prototyping* 19, no. 1 (2024): e2390504.
4. Ashok, Gaddam, Pankaj Kumar, and T. Ram Prabhu. "Global developments in additive manufacturing of polymer composite materials: A scientometric review." *Archives of Computational Methods in Engineering* (2025): 1–22.
5. Rajeshirke, Mithila, Ismail Fidan, Vivekanand Naikwadi, Suhas Alkunte, Ankit Gupta, and Mahdi Mohammadizadeh. "Material extrusion–based multi-material 3D printing: A holistic review of recent advances." *The International Journal of Advanced Manufacturing Technology* 139 (2025): 149–174.
6. Tahir, Muneeb, and Abdel-Fattah Seyam. "Greening fused deposition modeling: A critical review of plant fiber-reinforced PLA-based 3D-printed biocomposites." *Fibers* 13, no. 5 (2025): 64.
7. Sathiparan, Navaratnarajah, Pratheeba Jeyananthan, and Daniel Niruban Subramaniam. "A machine learning approach to predicting pervious concrete properties: A review." *Innovative Infrastructure Solutions* 10, no. 2 (2025): 55.
8. Cerchione, Roberto, Mariarosaria Morelli, Renato Passaro, and Ivana Quinto. "A critical analysis of the integration of life cycle methods and quantitative methods for sustainability assessment." *Corporate Social Responsibility and Environmental Management* 32, no. 2 (2025): 1508–1544.
9. Ng, Wai Lam, Azlin Mohd Azmi, Nofri Yenita Dahlan, and Kok Sin Woon. "Predicting life cycle carbon emission of green office buildings via an integrated LCA-MLR framework." *Energy and Buildings* 316 (2024): 114345.
10. Stojkovic, Miroslav, and Javaid Butt. "Industry 4.0 implementation framework for the composite manufacturing industry." *Journal of Composites Science* 6, no. 9 (2022): 258.
11. Bănică, Cristina-Florena, Alexandru Sover, and Daniel-Constantin Anghel. "Printing the future layer by layer: A comprehensive exploration of additive manufacturing in the era of Industry 4.0." *Applied Sciences* 14, no. 21 (2024): 9919.
12. Ali, S., I. Deiab, and S. Pervaiz. "State-of-the-art review on fused deposition modeling (FDM) for 3D printing of polymer blends and composites: Innovations, challenges, and applications." *The International Journal of Advanced Manufacturing Technology* 135 (2024): 5085–5113.

13. Vijayasankar, K. N., and Falguni Pati. "Effect of process parameters on the quality of additively manufactured PETG-silk composite." *Applied Composite Materials* 30, no. 1 (2023): 135–155.

14. Jankovic, Aleksandar, Gaurav Chaudhary, and Francesco Goia. "Designing the design of experiments (DOE)—An investigation on the influence of different factorial designs on the characterization of complex systems." *Energy and Buildings* 250 (2021): 111298.

15. Subramaniyan, Madheswaran, Sivakumar Karuppan, Prakash Eswaran, Anandhamoorthy Appusamy, and A. Naveen Shankar. "State of art on fusion deposition modeling machines process parameter optimization on composite materials." *Materials Today: Proceedings* 45 (2021): 820–827.

16. Li, Yalin, John T. Trimmer, Steven Hand, Xinyi Zhang, Katherine G. Chambers, Hannah AC Lohman, Rui Shi, Diana M. Byrne, Sherri M. Cook, and Jeremy S. Guest. "Quantitative sustainable design (QSD) for the prioritization of research, development, and deployment of technologies: a tutorial and review." *Environmental Science: Water Research & Technology* 8, no. 11 (2022): 2439–2465.

17. Zhu, Xingyu, Xianhai Meng, and Min Zhang. "Application of multiple criteria decision making methods in construction: A systematic literature review." *Journal of Civil Engineering and Management* 27, no. 6 (2021): 372–403.

Unravelling the Potential of Jute Fibre-Based Green Filament Composites via Fused Deposition Modelling

3

3.1 INTRODUCTION

The increasing environmental concerns and the global push towards sustainability have steered research communities and industries alike towards the development of green materials and eco-friendly manufacturing processes [1,2]. In this context, natural fibre-reinforced composites have gained remarkable attention due to their renewability, biodegradability, low density, and cost-effectiveness. Among various natural fibres, jute stands out as a promising candidate for polymer composite reinforcement, especially in

DOI: 10.1201/9781003732143-3

biodegradable matrices like polylactic acid (PLA). The present study seeks to pioneer the use of jute fibre-based green composite filaments specifically tailored for fused deposition modelling (FDM), a widely used additive manufacturing (AM) technology. Natural fibres such as jute have traditionally been used in a range of industries, including textiles, automotive interiors, packaging, and construction. However, their integration into 3D printing technologies—particularly in the form of printable composite filaments—is still in its nascent stages. This innovative convergence of natural materials and digital fabrication not only addresses the environmental impact of synthetic polymers and reinforcements but also enables novel applications through material customization. The effort to engineer jute-based composite filaments for FDM can be viewed as a foundational step in the development of sustainable digital manufacturing ecosystems.

3.1.1 Rationale for Selecting Jute Fibre

Jute is a lignocellulosic fibre composed primarily of cellulose, hemicellulose, and lignin. Its key advantages include high specific strength and stiffness, thermal resistance, fire retardancy, and availability in abundance, particularly in countries like India and Bangladesh [1]. Unlike synthetic fibres, jute is biodegradable and requires comparatively less energy for extraction and processing. While it may not surpass synthetic fibres like carbon or glass in mechanical performance, the eco-friendly characteristics of jute make it a preferable alternative for low-to-moderate-load applications in sustainable product design. What makes jute particularly suitable for integration into FDM filament systems is its compatibility with chemical surface modification, which significantly enhances interfacial adhesion with the polymer matrix [3,4]. In this work, semi-retted jute fibres are subjected to alkali treatment using sodium hydroxide (NaOH) to remove lignin and other amorphous components. This process improves surface roughness and exposes hydroxyl groups, thereby enhancing the compatibility with the PLA matrix. The fibres are then micronized and uniformly dispersed into PLA granules to produce composite feedstock for extrusion.

3.1.2 Bridging the Gap in Composite Fabrication

Traditionally, composite filaments are fabricated using twin-screw extruders, which allow for better mixing and dispersion of fillers in the polymer matrix.

However, such industrial-scale equipment is costly and often inaccessible to academic laboratories, small industries, and grassroots innovators. To address this, a low-cost, single-screw extrusion system was designed and developed for this study. The use of an auger bit in place of a conventional extrusion screw, coupled with precise temperature and speed control mechanisms via Arduino programming, enabled the effective fabrication of uniform composite filaments with diameters suitable for FDM (approximately 1.75 mm). This do-it-yourself (DIY) approach to filament extrusion not only democratizes the process of composite development but also paves the way for decentralized manufacturing. The ability to produce customized green filaments at low cost opens avenues for the reuse of agricultural residues, small-scale recycling, and educational purposes, thereby aligning with the principles of sustainability and circular economy.

3.1.3 Challenges and Opportunities in FDM Printing of Natural Fibre Composites

One of the core challenges in utilizing natural fibres in FDM is the heterogeneous nature of the feedstock. Differences in fibre size, moisture content, and interfacial properties can lead to nozzle clogging, uneven extrusion, and weak interlayer adhesion during printing. Additionally, PLA, though biodegradable and FDM-compatible, has limitations in thermal stability and mechanical durability, which can be exacerbated by poor filler dispersion or voids introduced during printing. Despite these challenges, successful printing of jute-reinforced PLA composites demonstrates the viability of such materials for prototyping and functional testing. Optimizing the printing parameters, such as nozzle temperature, bed temperature, infill density (ID), and print speed, is crucial to achieve acceptable mechanical performance [5,6]. In this study, systematic parametric studies and statistical modelling were employed to understand and control the effects of these variables on the tensile strength and emission profiles of the printed parts. The mechanical characterization of jute-PLA composites reveals that the addition of treated jute fibres improves the tensile strength compared to plasticized PLA alone, making them suitable for applications in packaging, consumer products, and structural components with moderate loading requirements. Moreover, the use of natural fibres contributes to lower emission rates of ultrafine particles (UFPs) and volatile organic compounds (VOCs) during printing, an important consideration for indoor manufacturing environments.

3.1.4 Contribution to Sustainable AM

The fabrication of jute-based composite filaments and their deployment in FDM marks an important contribution to the field of green AM. While much attention in AM has been given to improving mechanical strength, surface quality, and dimensional accuracy, relatively few studies focus on environmental performance indicators, such as emission reduction and life cycle impact. By combining experimental data with optimization algorithms such as particle swarm optimization (PSO) [7,8] and VIKOR (VlseKriterijumska Optimizacija I Kompromisno Resenje) [9], this research provides a multiobjective framework to balance performance and sustainability. The VIKOR method allows for decision-making under conflicting criteria, such as maximizing strength while minimizing emissions, by identifying the best trade-off solutions among multiple parameter combinations. On the other hand, PSO offers a stochastic approach to finding global optima in the parameter space, enabling adaptive learning from experimental outcomes. These modelling strategies enhance the robustness of material and process design, ultimately leading to more predictable and efficient printing outcomes.

3.2 CRAFTING THE FUTURE: PROCESSING AND MANUFACTURING OF JUTE/PLA COMPOSITES

The convergence of sustainable materials and advanced manufacturing technologies has catalyzed a transformation in how we perceive and produce engineering materials. Among the array of biodegradable composites, the amalgamation of jute fibre and PLA has drawn attention as an eco-conscious material system offering both environmental and mechanical value. Jute, a natural fibre abundant in lignocellulosic content, combined with PLA, a biodegradable thermoplastic derived from renewable resources like corn starch or sugarcane, represents a viable alternative to conventional composites derived from fossil-based synthetics. The present section outlines the methodology, challenges, and breakthroughs involved in processing and manufacturing jute/PLA composites using FDM—a technique poised to revolutionize green manufacturing.

3.2.1 Material Selection and Preparation: The Backbone of Green Composites

The first step in the development of jute/PLA composites involves the thoughtful selection of materials. PLA is preferred as the matrix material owing to its biodegradability, compatibility with FDM, ease of processing, and commendable mechanical properties. However, PLA on its own tends to be brittle and moisture-sensitive. To improve its mechanical strength and toughness, the reinforcement of natural fibres such as jute becomes an effective approach. Jute fibre, however, is not directly suitable for polymer matrix integration without pre-treatment. The untreated fibres contain surface impurities such as lignin, wax, oils, and hemicellulose, which hinder bonding with hydrophobic polymer matrices like PLA. Therefore, a chemical treatment process is employed to modify the jute surface [10,11]. In the current study, semi-retted jute was initially sourced from local suppliers and then subjected to alkali treatment using 5% NaOH solution. This process was conducted at 50°C for eight hours to eliminate the amorphous non-cellulosic materials. Periodic stirring ensured uniform exposure of the fibre surface to the chemical solution. After the alkali treatment, fibres were thoroughly washed with water and acetone to neutralize the residual alkali and then oven-dried at 70°C overnight. These steps enhanced fibre roughness, removed surface contaminants, and exposed hydroxyl functional groups, thereby improving interfacial adhesion with the PLA matrix. Subsequently, the fibres were mechanically cut and pulverized using a blender, and a 2-micron sieve was used to achieve a uniform short fibre size. This micron-scale fibre dimension is critical for enabling consistent melt flow and reducing the risk of nozzle blockage during the 3D printing process.

3.2.2 Filament Fabrication: From Raw Material to Printable Composite

A pivotal challenge in FDM-based composite fabrication is the development of compatible filaments. Unlike conventional extrusion-based composite manufacturing that uses industrial twin-screw extruders, this study proposes a low-cost, single-screw extrusion setup tailored for small-scale and lab-based applications. The extrusion setup was innovatively designed using an auger bit (300 mm in length and 16 mm in diameter) as a substitute for an extrusion screw, primarily to reduce costs and improve accessibility. The length-to-diameter ratio of 19:1 is comparable to commercial extrusion screws and ensures adequate shear mixing and residence time, leading to better blending of PLA and jute fibres. The extrusion module was equipped with a Nema 23 stepper motor, TB6600 driver, and an Arduino UNO-based control system to regulate the speed and direction of screw rotation. A k-type thermocouple and relay-controlled 35-watt heating element were incorporated to maintain a consistent temperature of around 185°C, sufficient to melt PLA without degrading the natural fibre reinforcement. The extrudate was manually pulled and cooled, producing filaments with diameters ranging from 1.5 to 1.8 mm, compatible with FDM nozzles. This desktop-compatible filament extruder has multiple advantages: it is low cost, portable, scalable, and flexible for various polymer-fibre combinations. It also presents opportunities for decentralized filament production using agricultural or household waste, aligning with the circular economy model.

3.2.3 Composite Sample Fabrication Using FDM

Once the filaments were fabricated and spooled, they were fed into a Creality Ender 3 FDM printer, a popular desktop-scale 3D printer. The processing parameters were carefully selected and varied during experimentation to understand their impact on mechanical and environmental performance.

Key FDM parameters included:

- **Nozzle diameter:** 1.5 mm
- **Layer height:** 0.3 mm
- **Infill density**: 50%, 75%, and 100%
- **Infill pattern:** Concentric
- **Printing temperature:** 180°C, 190°C, and 200°C
- **Build plate temperature:** 50°C, 60°C, and 70°C
- **Adhesion type:** Brim

Test specimens were printed as per ASTM D638 Type V [12] standards for tensile strength evaluation. The design and slicing were performed using open-source slicing software compatible with the printer, and necessary G-code adjustments were made to fine-tune bed levelling and extrusion control. A noteworthy challenge during printing was ensuring consistent filament feed, as even minor variations in diameter could disrupt flow or result in under-extrusion. Additionally, moisture absorption by jute fibres necessitated pre-print drying of the filaments to avoid porosity and steam bubbles within the print layers. Beyond mechanical performance, the environmental footprint of jute/PLA composite printing was assessed through emission studies. Measurements of Particle Number Concentrations (PNCs) and Specific Emission Rates (SERs) have been carried out.

3.2.4 Introduction to Parametric Analysis

In the context of AM, parametric analysis plays a crucial role in understanding how different process parameters influence the final properties of the printed parts. By systematically varying individual factors while holding others constant, the effects of main process variables on performance indicators such as tensile strength, print quality, or emission levels can be visualized and quantified. One of the primary tools for this analysis is the Main Effect Plot, which graphically represents the relationship between each input parameter and the output response [13]. For instance, plotting tensile strength against nozzle temperature can reveal whether strength increases linearly or non-linearly with temperature. Similarly, the effect of ID on mechanical performance or emissions can be visualized.

The main effect plots serve as diagnostic tools to:

- Identify which parameters have the most significant effect.
- Observe linear or non-linear trends.
- Help screen parameters for further modelling or optimization.

In this study, main effect plots were constructed to analyze the influence of:

- **Nozzle temperature (NT)** on tensile strength
- **Build plate temperature** on filament adhesion and surface finish
- **ID** on strength-to-weight ratio and material usage

The analysis provided initial insights into which parameters should be prioritized during optimization and served as the foundation for further statistical modelling.

3.2.5 Regression Analysis: Building Predictive Models

While parametric analysis is useful for visualization, it lacks predictive capability. To model the relationship between process parameters and output properties quantitatively, regression analysis was employed. Regression models are mathematical equations that relate one or more independent variables (e.g., temperature, ID) to a dependent variable (e.g., tensile strength or emission rate).

In this study, multiple linear regression (MLR) was used to generate equations of the form:

$$Y = \beta_0 + \beta_1 X_1 + \beta_2 X_2 + \beta_3 X_3 + \cdots + \epsilon$$

where:

- Y: Output response (e.g., tensile strength)
- X_1, X_2, X_3: Input parameters (e.g., temperature, ID)
- β: Regression coefficients
- ϵ: Random error term

The regression coefficients were estimated using least-squares fitting to the experimental data. High R^2 values (coefficient of determination) indicated that the models fit the data well and could be used to predict outputs for unseen parameter combinations.

Importantly, the regression models enabled:

- **Sensitivity analysis:** Understanding which parameters have the most influence.
- **Interaction detection:** Observing whether the effect of one parameter depends on the level of another.
- **Objective function formulation:** Used later in optimization techniques like **VIKOR** and **PSO**.

3.2.6 Statistical Modelling and Optimization of Process Parameters

A novel feature of this study was the dual-objective optimization of the FDM process—maximizing tensile strength while minimizing particulate emissions such as UFPs and VOCs, which are often neglected in performance assessments. To achieve this, VIKOR and PSO techniques were employed.

VIKOR Method: This multi-criteria decision-making (MCDM) approach helps identify a compromise solution that provides the closest-to-ideal performance under multiple conflicting objectives. For this study:

- Inputs: Printing temperature, bed temperature, ID
- Outputs: Tensile strength (to be maximized), emission levels (to be minimized)

Ideal and anti-ideal solutions were determined based on experimental data. Utility and regret measures were calculated for each combination, followed by computing the VIKOR index, which ranked the printing parameter sets in ascending order of desirability.

PSO Algorithm: PSO was used as a metaheuristic optimizer to explore the search space of parameters. Each "particle" represented a potential solution (a combination of printing temperature and bed temperature), and its fitness was evaluated using regression-based objective functions derived from empirical data.

Particles updated their velocities and positions iteratively using:

- Inertia weight for balancing exploration and exploitation
- Local (pbest) and global (gbest) optima tracking
- Random perturbation factors to prevent premature convergence

Through these iterative steps, optimal settings for low emissions and high mechanical performance were derived. This approach underscores the importance of modelling both functional and environmental outputs for sustainable manufacturing.

3.3 PROBING THE ESSENCE: PROPERTY EVALUATION OF FDM-PRINTED JUTE/PLA COMPOSITE

As sustainability becomes a central design and manufacturing criterion, the development of natural fibre-reinforced biodegradable composites such as jute/PLA has gained considerable traction. However, for such materials to be viable alternatives to conventional petroleum-based composites, it is essential to evaluate their mechanical performance, thermal behaviour, and environmental safety, particularly when fabricated using FDM. This section presents a comprehensive property evaluation of FDM-printed jute/PLA composites, with a focus on their mechanical behaviour, fractographic features, and environmental emissions. The findings are substantiated with tabular data, figures, and statistical analysis to provide a nuanced understanding of material performance.

3.3.1 Mechanical Testing: Unveiling Performance under Load

The tensile strength and elongation at break of jute/PLA specimens were tested according to ASTM D638 Type V standards using a Universal Testing Machine

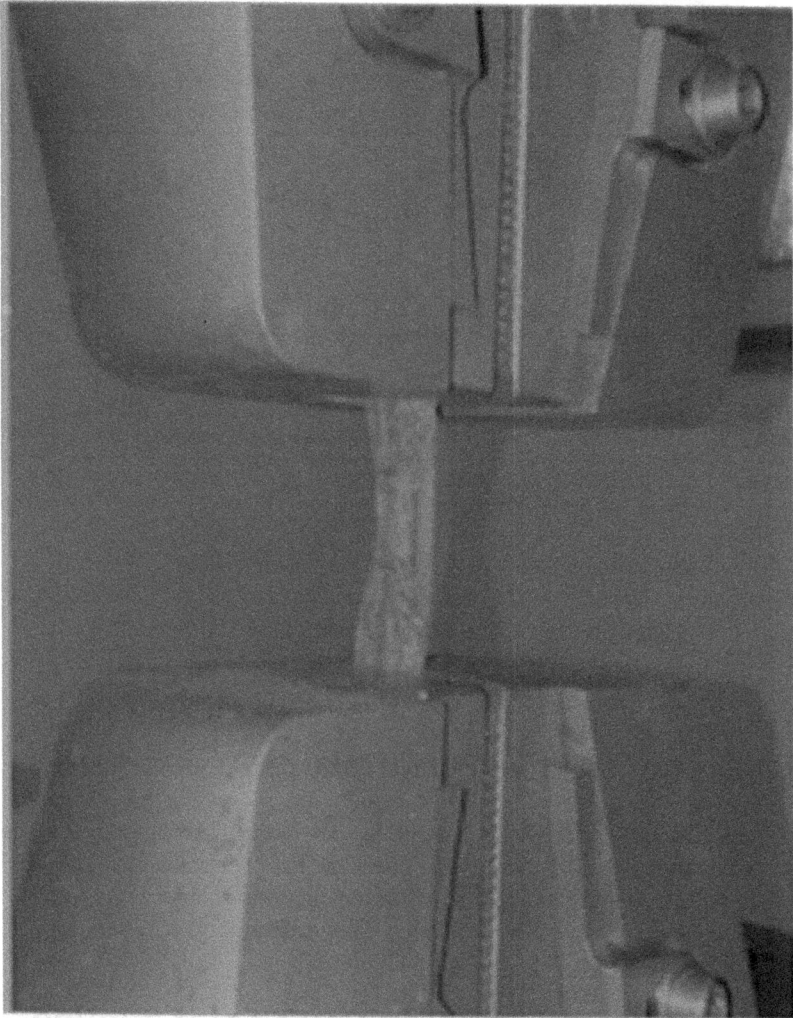

FIGURE 3.1   Sample gripped in the Universal Testing Machine before applying load with gauge length marked

(Figure 3.1). Specimens were fabricated at varying combinations of nozzle temperature, bed temperature, and ID. Table 3.1 summarizes the key results.

The sample with 200°C nozzle temperature and 100% infill (S3) achieved the highest ultimate tensile strength (UTS) of 47.5 MPa. Increased ID and optimal extrusion temperature contributed significantly to performance improvement.

TABLE 3.1 Tensile properties of FDM-printed jute/PLA composites

SAMPLE ID	NOZZLE TEMP (°C)	BED TEMP (°C)	INFILL DENSITY (%)	UTS (MPA)	ELONGATION AT BREAK (%)
S1	180	50	50	38.2	2.1
S2	190	60	75	44.6	2.6
S3	200	70	100	47.5	3.1
S4	190	60	100	46.1	2.9
S5	180	70	75	42.0	2.3

Abbreviations: FDM, fused deposition modelling; PLA, polylactic acid; UTS, ultimate tensile strength.

TABLE 3.2 Comparative tensile strength of PLA-based composites

COMPOSITE TYPE	REINFORCEMENT	PROCESS	UTS (MPA)
PLA (neat)	None	FDM	36.0
PLA + Wood	Wood fibre	FDM	40.0
PLA + Carbon Fibre	Chopped carbon fibre	FDM	26.6
PLA + Jute (Current study)	Chemically treated jute	FDM	47.5

3.3.2 Comparative Performance Assessment

To contextualize performance, jute/PLA composites were compared with literature-reported FDM composites. The same has been shown in Table 3.2.

Jute/PLA composites show superior tensile performance among FDM-compatible green composites, validating their structural capability.

3.3.3 Emission Profile: Towards Safer Printing

To ensure environmental and health safety, emissions during FDM printing were measured. Two primary metrics were assessed:

- PNC
- SER [14] presented in Table 3.3

TABLE 3.3   Emission results from FDM printing

MATERIAL	PNC (PARTICLES/CM³)	SER (µG/H)
PLA	2.1×10^{10}	430
ABS	1.9×10^{11}	780
PLA + Carbon	1.6×10^{11}	710
PLA + Jute	$\mathbf{8.2 \times 10^{9}}$	**295**

Abbreviations: ABS, acrylonitrile butadiene styrene; PNC, particle number concentration; SER, specific emission rate.

TABLE 3.4   Summary of evaluated properties

PROPERTY TYPE	RESULT/OBSERVATION
Tensile strength	Max: 47.5 MPa at 200°C nozzle and 100% infill
Ductility	Elongation at break improved with higher temperatures
Emission performance	PNC reduced by >50% compared to ABS/CF-PLA

Abbreviation: CF, carbon fibre.

Jute/PLA composites released less than half the emissions compared to carbon-PLA and acrylonitrile butadiene styrene (ABS), making them ideal for indoor 3D printing setups such as labs and offices.

3.3.4 Durability and Environmental Resistance

Although not tested extensively in this study, initial findings and literature suggest that:

- **Alkali-treated jute** improves water resistance slightly but remains hydrophilic.
- **Moisture exposure** may reduce interfacial bonding over time.
- **Further research** is needed on ultraviolet stability, impact resistance, and long-term durability.

Recommendation: Incorporating hydrophobic agents or coatings can enhance outdoor performance for applications in construction or agriculture. The summary of the evaluated properties has been presented in Table 3.4.

3.4 STATISTICAL ALCHEMY: ANALYZING FDM-PRINTED JUTE/PLA COMPOSITE

The transformation of raw data into knowledge, and knowledge into engineering wisdom, lies at the heart of what can be termed "statistical alchemy," the art and science of drawing meaningful insights from experimental results. In the context of FDM of jute fibre-reinforced PLA composites, statistical analysis plays a pivotal role in not just understanding how various processing parameters influence performance, but also in identifying the optimal configurations that maximize mechanical strength while minimizing emissions and manufacturing complexity. This subsection explores the application of statistical techniques for modelling, analyzing, and optimizing the jute/PLA FDM process. It begins with experimental design strategies, proceeds to parametric and regression analysis, and culminates with multi-objective decision-making using VIKOR and PSO. These tools collectively help decode the hidden structure behind the material–process–performance relationships in green composite printing.

3.4.1 Experimental Design and Data Collection Strategy

To study the effects of processing parameters on performance, a design of experiments approach was adopted. The selection of input variables was guided by prior literature and preliminary trials, with three key parameters considered for this study:

- **Nozzle Temperature (NT):** 180°C, 190°C, 200°C
- **Bed Temperature (BT):** 50°C, 60°C, 70°C
- **Infill Density (ID):** 50%, 75%, 100%

Using a full factorial design ($3^3 = 27$ combinations) would have been exhaustive and resource-intensive. Hence, a Taguchi L_9 orthogonal array was chosen to reduce the experimental load while still capturing the main effects and some interactions. Each of the nine experiments resulted in printed ASTM D638 Type V tensile specimens, whose UTS and emission rate (PNC) were recorded as response variables.

3.4.2 Parametric Analysis and Main Effect Interpretation

The main effect plots are graphical tools that show how each input parameter individually affects the response. These plots were used to quickly identify which parameters had the strongest influence on tensile strength and emissions (Figure 3.2).

Insights:

- **NT:** Increasing NT from 180°C to 200°C improves interlayer fusion, resulting in higher strength.
- **ID:** As ID increases, more material supports the load, leading to better tensile behaviour.
- **BT:** Affects adhesion and thermal shrinkage; its effect is non-linear.

Optimization must balance tensile performance and emission control—higher temperatures improve strength but worsen emissions.

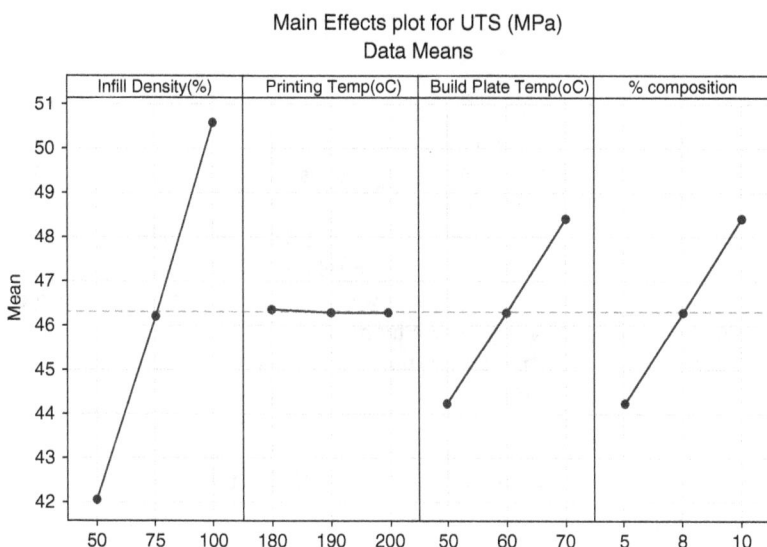

FIGURE 3.2 Nozzle temperature and infill density exhibit strong influence on ultimate tensile strength, while bed temperature has a moderate effect

TABLE 3.5   Model performance metrics

METRIC	VALUE
R^2 (Coefficient of determination)	0.93
Adjusted R^2	0.91
RMSE	1.8 MPa
p-Values	<0.05 for all predictors

Abbreviations: RMSE, root mean square error.

3.4.3 MLR Modelling

To create a predictive model for UTS based on the three input parameters, an MLR equation [15] was developed:

$$UTS = \beta_0 + \beta_1 \cdot NT + \beta_2 \cdot BT + \beta_3 \cdot ID$$

The coefficients were determined via least-squares estimation using the L_9 dataset:

$$UTS = -20.45 + 0.32 \cdot NT + 0.25 \cdot BT + 0.48 \cdot ID$$

where:

- NT: Nozzle temperature (in °C)
- BT: Bed temperature (in °C)
- ID: Infill density (in %)

A summary of the model performance metrics is shown in Table 3.5.

The model has high predictive power and reveals that ID is the most influential factor, followed by nozzle temperature.

A similar regression model was built for emissions (PNC), which showed that nozzle temperature had the strongest influence on particle release.

3.4.4 Multi-Objective Optimization: VIKOR Approach

The experimental problem involved conflicting objectives: maximizing tensile strength (UTS) while minimizing emissions (PNC). The VIKOR method, an MCDM tool, was employed. The decision matrix is shown in Table 3.6.

The VIKOR method identified R4 (190°C, 60°C, 100%) as the best compromise, offering high strength with relatively low emissions.

TABLE 3.6   VIKOR decision matrix

RUN	NT (°C)	BT (°C)	ID (%)	UTS (MPA)	PNC (×10⁹)	RANK
R1	180	50	50	38.2	8.4	4
R2	190	60	75	44.6	7.2	3
R3	200	70	100	47.5	9.1	2
R4	190	60	100	46.1	6.9	1

Abbreviations: BT, bed temperature; ID, infill density; NT, nozzle temperature; VIKOR, VlseKriterijumska Optimizacija I Kompromisno Resenje.

3.4.5 Optimization via PSO

To complement the MCDM analysis, PSO was implemented using the regression models as fitness functions.

3.4.5.1 Objective Functions

- f_1 = UTS (to maximize)
- f_2 = PNC (to minimize)

Using a swarm size of 20 particles over 100 iterations, the PSO algorithm converged to:

- **Optimal NT:** 192°C
- **Optimal BT:** 62°C
- **Optimal ID:** 98%

PSO reinforced the VIKOR findings, adding robustness to the selection of printing parameters.

3.4.6 Validation and Error Analysis

Experimental runs using PSO-optimized parameters yielded:

- **Measured UTS:** 46.8 MPa
- **Predicted UTS:** 47.1 MPa
- **Prediction Error:** 0.64%

This small error validates the regression model and optimization outcomes.

TABLE 3.7 Statistical takeaways

TOOL USED	PURPOSE	OUTCOME
DOE (Taguchi)	Experimental simplification	Efficient data generation
Main Effect	Factor screening	Identified dominant variables
Regression	Predictive modelling	$R^2 = 0.93$
VIKOR	Trade-off optimization	Best compromise solution
PSO	Global optimization	Verified and fine-tuned output

Abbreviations: DOE, design of experiments; PSO, particle swarm optimization.

3.4.7 Statistical Synthesis and Engineering Relevance

The integration of parametric analysis, regression modelling, and multi-objective optimization provides an end-to-end strategy for refining the FDM process of green composites. Key outcomes include:

- Identification of dominant parameters (ID > NT > BT)
- Development of high-fidelity predictive models
- Use of decision-making tools (VIKOR, PSO) for balancing performance and sustainability
- Empirical validation of optimal parameter settings

This statistical ecosystem enables knowledge-driven AM, reducing trial-and-error and fostering reproducibility in sustainable materials processing.

Statistical alchemy, the conversion of empirical data into predictive power, has been successfully demonstrated in this study. By applying design of experiments, regression analysis, and optimization, the FDM process of jute/PLA composites has been systematically decoded and refined. The summary of the statistical takeaways has been tabulated in Table 3.7.

3.5 FORGING AHEAD: CHARTING THE FUTURE OF JUTE FIBRE-BASED GREEN FILAMENT COMPOSITES

The increasing environmental consciousness and global regulatory pressure on non-renewable and non-degradable materials have triggered an irreversible momentum towards sustainable manufacturing. At the confluence of this green transformation lies an opportunity to explore the next frontier of materials: jute fibre-based green filament composites fabricated via FDM. The journey so far has shown promising results in material development, process optimization, and mechanical performance evaluation. However, to fully harness the environmental and functional potential of these materials, a forward-looking roadmap must be charted. This section aims to outline the strategic directions, challenges, and research priorities that can drive the future of jute fibre-reinforced PLA (or other biopolymers) in the context of AM. By identifying gaps in current knowledge and proposing integrated, interdisciplinary pathways, this discourse envisions how jute-based composites can evolve from being academic curiosities to industrially viable, environmentally friendly solutions.

3.5.1 Consolidating Current Progress

Before forging into the future, it is essential to reflect upon the major milestones that have enabled jute-based filament composites to emerge as credible green alternatives:

- **Chemical Surface Modification:** The use of alkali treatment, silane coupling, or compatibilizers to enhance fibre-matrix adhesion has significantly improved the mechanical performance of jute/PLA composites.
- **DIY Filament Extrusion:** Low-cost, desktop-scale filament extruders have democratized the fabrication process, enabling material research in decentralized environments.
- **Mechanical Viability:** With UTS values reaching ~47 MPa, these composites are now suitable for low-to-moderate-load applications in sectors like packaging, consumer goods, and prototyping.
- **Emission Reduction:** Compared to synthetic fibre-reinforced composites, jute-based systems emit lower levels of UFPs and VOCs, making them safer for indoor and office-based printing.

However, the above achievements are foundational. Much remains to be done before these materials can be implemented at scale, integrated into circular economy frameworks, and adapted to advanced functional applications.

3.5.2 Broadening Material Horizons

Currently, most jute-based composite studies focus on PLA as the base matrix due to its biodegradability, thermal compatibility with jute, and commercial availability. Moving forward, other bio-based or biodegradable polymers should be explored to expand the material scope:

1. Polyhydroxyalkanoates:
   - Derived from microbial fermentation.
   - Superior barrier properties compared to PLA.
   - Better environmental degradation rate, though more expensive.
2. Thermoplastic Starch:
   - Abundant and inexpensive.
   - Moisture-sensitive and requires blending or plasticization.
3. Polybutylene Succinate and Polycaprolactone:
   - Good flexibility and thermal properties.
   - Compatible with jute but need further evaluation in FDM contexts.

By evaluating the mechanical, thermal, and environmental compatibilities of these matrices with jute, researchers can build a catalogue of green composite filaments optimized for specific use cases (e.g., packaging vs. structural parts).

3.5.3 Advanced Fibre Treatments and Hybridization

While alkali treatment improves jute fibre properties, advanced fibre treatments can open new avenues for performance enhancement:

- **Enzymatic Treatment:** Environmentally benign method to improve surface roughness and remove lignin.
- **Plasma or Corona Discharge:** Physical methods to enhance surface energy and functional group availability.
- **Grafting Techniques:** Introduction of functional groups onto the fibre surface to improve chemical affinity with matrix polymers.

Additionally, hybrid composites that combine jute with other natural or synthetic fibres (e.g., bamboo, sisal, basalt, or carbon) can balance biodegradability and performance:

- **Jute-Carbon Hybrids:** Good strength with reduced environmental cost.
- **Jute-Basalt Composites:** Higher fire resistance and mechanical strength.

The use of *nano-fillers* (e.g., cellulose nanocrystals, graphene oxide) can further tailor thermal stability, electrical conductivity, and mechanical integrity.

3.5.4 Life Cycle Assessment and Circularity

Sustainability is incomplete without a full life cycle assessment. Future work should aim to quantify:

- **Carbon footprint** of jute/PLA production vs. ABS, polyethylene terephthalate glycol, or carbon fibre-PLA.
- **End-of-life scenarios:** Composability, recyclability, biodegradation in soil/marine environments.
- **Energy inputs** during filament extrusion and printing.

Moreover, integration of jute waste from agriculture or textile industries into composite production can enhance circularity. Using locally available fibres in decentralized filament fabrication could also reduce transportation emissions and support rural livelihoods.

3.5.5 Application-Specific Design and Functionalization

To make these materials more than substitutes, they must be functionally superior or application-specific. Promising application domains include:

1. Biomedical Devices:
   - Customized, biodegradable surgical tools or supports.
   - PLA and jute are non-toxic and can be further functionalized for antimicrobial properties.
2. Smart Packaging:
   - Natural look and feel for premium eco-packaging.
   - Integration of moisture indicators or biodegradable sensors.
3. Interior Decor and Architecture:
   - Lightweight panels, customized tiles, or room dividers.
   - Jute offers aesthetic appeal, acoustic dampening, and low thermal conductivity.
4. Automotive Interiors:
   - Already used in injection moulding; FDM can enable customization and repairability.

Functionalization through coating (e.g., beeswax, biodegradable polymers) or incorporation of additives (e.g., flame retardants, anti-fungal agents) can diversify the applications significantly.

3.5.6 Process Innovation and Hardware Integration

In terms of fabrication, further research can enhance process repeatability, hardware compatibility, and scalability:

- **Twin-Screw Extrusion at Lab Scale:** Better fibre dispersion and throughput.
- **Automated Drying Systems:** Reduce moisture during filament storage and feeding.
- **Printer Modifications:** Hardened nozzles, flexible filament feeders, and real-time temperature control for better composite print quality.
- **Closed-Loop Monitoring:** Use of thermal cameras, load cells, or acoustic sensors to monitor defects or warping during print runs.

These enhancements can turn desktop FDM setups into micro-factories capable of producing functional eco-products.

3.5.7 Digital Twin and Artificial Intelligence-Driven Optimization

With growing computational capabilities, the use of digital twins and artificial intelligence (AI) in AM is gaining traction. In the context of jute/PLA composites:

- **Digital Twins:** Create virtual replicas of material and process models to simulate, predict, and optimize print behaviour.
- **Machine Learning Models:** Train regression or classification models to predict mechanical properties or failure probability from process parameters and environmental conditions.
- **Multi-Objective Optimization:** Real-time adjustment of parameters to balance strength, emissions, and print time.

Such approaches reduce experimental loads, improve reproducibility, and enable mass customization with green materials.

3.5.8 Policy Integration and Standardization

For wide-scale adoption, certification and policy frameworks must evolve:

- **Bio-Based Material Certification:** Validate carbon savings, toxicity, and degradation profiles.
- **FDM Composite Standards:** Develop ASTM/ISO standards for testing and benchmarking jute-based filaments.
- **Incentivizing Local Production:** Policies that support rural industries or maker spaces using natural fibres.

Open-source platforms and repositories could host formulation recipes, process guidelines, and performance databases for easy global dissemination.

3.5.9 Challenges to Overcome

Despite its promise, several challenges still inhibit the full-scale deployment of jute-based filament composites and the same is showcased in Table 3.8.

Addressing these issues will require multidisciplinary collaboration across materials science, mechanical engineering, sustainability studies, and data science.

TABLE 3.8 Challenges that inhibit the full-scale deployment of jute-based filament composites

CHALLENGE	IMPACT	POTENTIAL SOLUTION
Fibre moisture sensitivity	Leads to porosity, reduced strength	Pre-print drying, moisture-resistant coatings
Irregular filament diameter	Causes nozzle clogging or poor flow	Real-time diameter sensors, advanced extrusion control
Limited interfacial compatibility	Causes fibre pull-out, low load transfer	Surface treatment, compatibilizers
Lack of long-term durability data	Unclear performance under UV, moisture, fatigue	Accelerated aging studies
Lack of design guidelines	Hinders industrial adoption	FDM-specific design handbooks for green composites

Abbreviations: FDM, fused deposition modelling.

3.5.10 Conclusion: Towards a Greener Future with AM

The road ahead for jute fibre-based green filament composites is full of possibilities. By integrating materials innovation, process optimization, and digital tools, these bio-composites can transition from labs to real-world applications with high environmental, social, and economic impact.

3.5.10.1 Strategic Research Priorities Moving Forward

1. Expand the matrix base beyond PLA to other biodegradable polymers.
2. Develop hybrid and nano-enhanced jute composites for multi-functionality.
3. Create AI-augmented digital twins for process control and defect prediction.
4. Conduct thorough life cycle and techno-economic assessments.
5. Build open-source platforms for design, formulation, and data sharing.

AM offers the flexibility, decentralization, and customization capabilities needed to empower local communities, reduce environmental burden, and fuel the next generation of eco-conscious manufacturing. With materials like jute leading the way, the future is not just printable—it's sustainable.

3.6 CONCLUSIONS

The exploration of jute fibre-based green filament composites through the lens of FDM marks a pivotal step in the sustainable transformation of AM. This chapter has comprehensively examined the potential of jute/PLA composites, from material formulation and filament processing to property evaluation, statistical modelling, and strategic forecasting. The findings underscore the viability of using natural, biodegradable resources to develop functional, eco-friendly alternatives to conventional petroleum-derived materials in 3D printing.

Key insights reveal that:

- Chemically treated jute fibres, when reinforced in PLA and processed through FDM, offer competitive mechanical performance with significantly reduced environmental impact.

- Optimization of printing parameters such as nozzle temperature, bed temperature, and ID plays a critical role in balancing structural integrity and emission control.
- Statistical techniques including main effect plots, regression modelling, VIKOR, and PSO enable data-driven decision-making, maximizing the composite's utility and safety.
- Prospects are promising, with pathways emerging for new matrix systems, hybrid reinforcement, functional applications, and AI-integrated smart manufacturing platforms.

While challenges such as moisture sensitivity, filament consistency, and long-term durability persist, they are surmountable with continued interdisciplinary research. The integration of materials science, digital manufacturing, environmental engineering, and data analytics offers a compelling framework for refining and scaling this green technology. Ultimately, jute fibre-based FDM composites embody more than just material innovation, they represent a paradigm shift towards localized, circular, and sustainable manufacturing ecosystems. As we move towards a future defined by ecological responsibility and technological agility, such innovations will serve as cornerstones in the journey towards greener industry practices.

REFERENCES

1. Abdur Rahman, M., Serajul Haque, Muthu Manokar Athikesavan, and Mohamed Bak Kamaludeen. "A review of environmental friendly green composites: Production methods, current progresses, and challenges." *Environmental Science and Pollution Research* 30, no. 7 (2023): 16905–16929.
2. Yadav, Mithilesh, Anil Kumar Maurya, and Sujeet Kumar Chaurasia. "Eco-friendly polymer composites for green packaging: Environmental policy, governance and legislation." In *Sustainable Packaging Strengthened by Biomass*, edited by Arbind Prasad, J. Paulo Davim, Sonika Gupta, and Sushil Kumar Verma, pp. 317–346. Woodhead Publishing, 2025.
3. Devarajan, Balaji, Rajeshkumar LakshmiNarasimhan, Bhuvaneswari Venkateswaran, Sanjay Mavinkere Rangappa, and Suchart Siengchin. "Additive manufacturing of jute fiber reinforced polymer composites: A concise review of material forms and methods." *Polymer Composites* 43, no. 10 (2022): 6735–6748.
4. Demir, Murat, and Yasemin Seki. "Interfacial adhesion strength between FDM-printed PLA parts and surface-treated cellulosic-woven fabrics." *Rapid Prototyping Journal* 29, no. 6 (2023): 1166–1174.
5. Popović, Mihajlo, Miloš Pjević, Aleksa Milovanović, Goran Mladenović, and Miloš Milošević. "Printing parameter optimization of PLA material

concerning geometrical accuracy and tensile properties relative to FDM process productivity." *Journal of Mechanical Science and Technology* 37, no. 2 (2023): 697–706.

6. Khudhair, Mustafa T., and Kareem N. Salloomi. "Parametric optimization of layer thickness, speed, and high acceleration on surface roughness, productivity, and quality of 3D printed PLA objects." *Progress in Additive Manufacturing* 10 (2025): 9775–9792.

7. Wang, Feng, Heng Zhang, and Aimin Zhou. "A particle swarm optimization algorithm for mixed-variable optimization problems." *Swarm and Evolutionary Computation* 60 (2021): 100808.

8. Yao, Jicheng, Xiaonan Luo, Fang Li, Ji Li, Jundi Dou, and Hongtai Luo. "Research on hybrid strategy Particle Swarm Optimization algorithm and its applications." *Scientific Reports* 14, no. 1 (2024): 24928.

9. Pan, Rongshun, Jiahao Yu, and Yongman Zhao. "Many-objective optimization and decision-making method for selective assembly of complex mechanical products based on improved NSGA-III and VIKOR." *Processes* 10, no. 1 (2021): 34.

10. Boey, Jet Yin, Siti Baidurah Yusoff, and Guan Seng Tay. "A review on the enhancement of composite's interface properties through biological treatment of natural fibre/lignocellulosic material." *Polymers and Polymer Composites* 30 (2022): 09673911221103600.

11. Mohammadi, Majid, Mohamad Ridzwan Ishak, and Mohamed Thariq Hameed Sultan. "Exploring chemical and physical advancements in surface modification techniques of natural fiber reinforced composite: A comprehensive review." *Journal of Natural Fibers* 21, no. 1 (2024): 2408633.

12. ASTM Subcommittee D20. 10 on Mechanical Properties. "Standard test method for tensile properties of plastics." *American Society for Testing and Materials* (1998).

13. Okoro, Oseweuba Valentine, D. E. Caevel Hippolyte, Lei Nie, Keikhosro Karimi, Joeri F. M. Denayer, and Armin Shavandi. "Machine learning-based predictive modeling and optimization: Artificial neural network-genetic algorithm vs. response surface methodology for black soldier fly (Hermetia illucens) farm waste fermentation." *Biochemical Engineering Journal* 218 (2025): 109685.

14. Hejna, Aleksander, Mariusz Marć, Paweł Szymański, Kamila Mizera, and Mateusz Barczewski. "Analysis of emission of volatile organic compounds and thermal degradation in investment casting using fused deposition modeling (FDM) and three-dimensional printing (3DP) made of various thermoplastic polymers." *Environmental Science and Pollution Research* 31, no. 50 (2024): 60371–60388.

15. Singh, Priyanka, Abiola Adebanjo, Nasir Shafiq, Siti Nooriza Abd Razak, Vicky Kumar, Syed Ahmad Farhan, Ifeoluwa Adebanjo, et al. "Development of performance-based models for green concrete using multiple linear regression and artificial neural network." *International Journal on Interactive Design and Manufacturing (IJIDeM)* 18, no. 5 (2024): 2945–2956.

Unleashing the Potential of FDM-Printed Coir-Based Green Composites

4

4.1 INTRODUCTION

Rapid prototyping (RP) is an emerging manufacturing process that caters to small-scale industry applications with minimal material wastage and reduced inventory requirements. Among the various RP techniques, fused deposition modelling (FDM) is considered one of the most cost-effective methods for small-scale vendors [1]. In FDM, materials such as metals, ceramics, and polymers are used as feedstock. However, polymers are most commonly utilized in small-scale industries, including applications like food packaging, drug delivery, and children's toys. Generally, polymers are classified into two main categories: thermosetting polymers and thermoplastics (TPs). Thermosetting polymers such as Bakelite, epoxy resins, melamine, and vulcanized rubber are widely used in various industries due to their excellent properties. In some applications, these materials have even replaced metals. However, thermosetting polymers have certain drawbacks, such as irreversibly hardening when heated, making them impossible to reshape or reuse [2].

DOI: 10.1201/9781003732143-4

Additionally, they are non-biodegradable. Therefore, there is a need for new polymers that can overcome these limitations.

TPs exhibit superior properties and address many of the shortcomings associated with thermosetting polymers. Among the various TPs, polylactic acid (PLA) is widely used due to its biodegradability, eco-friendliness, and renewable origin, as it is derived from starch-based materials [3]. Eldeeb et al. [4] investigated the mechanical properties of PLA components fabricated using FDM. Their findings revealed that the mechanical properties of the PLA samples were significantly affected by the FDM process parameters. Chalgham et al. [5] conducted a post-thermal analysis of PLA samples printed using the FDM process. The results showed that the optimal process parameter settings were a nozzle temperature (NT) of 190°C, a layer thickness (LT) of 0.3 mm, and a print speed of 90 mm/s, based on the maximum bending force as the response parameter. Furthermore, samples prepared with these settings and subsequently thermally treated at 75°C exhibited a significant increase in bending force compared to those tested before thermal treatment.

To enhance the mechanical properties of PLA components printed by FDM, the optimization of FDM process parameters plays a crucial role. Shahrjerdi et al. [6] investigated the optimal process parameter settings using response surface methodology (RSM) and statistical modelling analysis to improve the mechanical properties of PLA components. The results indicated that the optimal parameter settings were an NT of 220°C, an infill density (ID) of 100%, and an LT of 0.3 mm. Additionally, Patil et al. [7] studied the improvement of the mechanical properties of PLA samples by determining the optimal process parameters using RSM during FDM printing. The results revealed that the optimal parameter settings i.e., an LT of 0.2 mm, an ID of 80%, and a build orientation of 60° are significantly enhanced the mechanical properties of the PLA samples.

However, PLA alone exhibits lower mechanical strength, thermal stability, and biodegradability. As a result, limited research has been conducted on the characterization of PLA reinforced with natural fibre samples printed using FDM. Kacem et al. [8] investigated the mechanical properties of PLA reinforced with yucca fibre samples fabricated through FDM. The results showed that the PLA composite samples exhibited improvements of 22% in tensile strength, 20% in compressive strength, and 12% in flexural strength compared to virgin PLA. Wang et al. [9] fabricated PLA-based samples reinforced with ramie using FDM and studied their mechanical properties. The results showed that the continuous ramie yarn enhanced both the tensile strength and modulus of the samples compared to pure PLA. Additionally, research has been conducted on the mechanical properties of PLA polymer reinforced with sugarcane bagasse [10], hemp fibre [11], wood fibre [12], bamboo and flax fibres [13], and rice husk [14], with samples fabricated using FDM and compared to pure

PLA samples. However, the optimization of FDM process parameters remains a bottleneck issue in the printing of PLA composite samples.

Patil et al. [15] studied the optimization of FDM process parameters for printing PLA samples reinforced with rice husk fibres. The results showed that the mechanical properties were enhanced at the optimal process parameter settings, namely an LT of 0.2 mm, an NT of 190°C, and 1 wt% rice husk content. Natarajan et al. [16] investigated the mechanical properties of PLA reinforced with Acacia concinna fibre samples fabricated using FDM. The findings revealed that LT and ID were the most significant process parameters influencing the mechanical performance of the composite samples.

However, limited research has been conducted on the optimization of FDM process parameters for PLA samples reinforced with natural fibres and additives. This chapter explores the potential of FDM-printed coir-based green composites. Initially, the coir fibres were processed and chemically treated with an NaOH solution to enhance their hydrophobic properties. Subsequently, PLA reinforced with coir fibres and additives was used to produce filaments via a filament extrusion setup, which served as the feedstock for FDM. Tensile test samples of the PLA composites were then fabricated using FDM and subjected to mechanical testing. Finally, statistical model analysis was conducted to evaluate the significance of independent process parameters on the response variables using analysis of variance (ANOVA) and regression analysis. The optimal process parameter settings were identified under conditions that yielded the maximum values of the response parameters.

4.2 MATERIALS AND METHODS

4.2.1 Bio Polymers and Coupling Agents

PLA pellets, sourced from Konkan Specialty Poly Products Pvt. Ltd., Mangalore, were selected as the primary biodegradable polymer matrix due to their excellent printability and low environmental impact. To enhance the mechanical properties of the biodegradable PLA, natural lignocellulosic coir fibres were incorporated. Additives such as polyethylene glycol and maleic anhydride were used to improve the ductility of the matrix, facilitate flowability during extrusion, and enhance interfacial adhesion between the hydrophobic PLA matrix and the hydrophilic coir fibres [17]. Additionally, the coir fibres underwent NaOH treatment to remove unwanted chemical constituents and to clean their surface, thereby improving fibre-matrix bonding [18].

4.2.2 Processing of Coir Fibres

Coir fibres were extracted from dried coconut shells using a peeling process. The extracted long fibres were then dried in a hot air oven at 60°C for 24 hours to eliminate any moisture content. These long fibres were subsequently cut into smaller pieces and ground into fine coir fibres with an average size of 300 μm. A sieving process was carried out to ensure uniformity in the size of the small coir fibres (SCFs). The SCFs were then treated with a 5 wt% NaOH solution at 50°C for four hours. After treatment, the fibres were removed from the NaOH solution and washed with an acetone solution to eliminate any unreacted NaOH [19]. The fibres were then thoroughly rinsed with deionized water until the desired pH level was achieved. Finally, the wet, treated SCFs were dried in a hot air oven at 60°C for 24 hours. This NaOH treatment effectively removes unwanted chemical constituents, cleans the fibre surface of any impurities, and enhances the hydrophobic nature of the fibres by increasing surface roughness [18]. Figure 4.1 shows the step-by-step procedure of processing of coir fibres.

FIGURE 4.1   Processing of coir fibres and chemical treatment

4.2.3 Preparation of PLA Composite Samples by FDM Process

The blended compound of PLA pellets, SCFs, and additives was produced using a multi-screw melting mixer. The resulting blend was chopped into small granules through a cutting process and then fed into a filament extrusion device to produce PLA composite filament with a uniform and continuous diameter of 1.75 mm. This PLA-based filament, reinforced with SCFs and additives, was used as the feedstock material for FDM. Tensile samples of both pure PLA and PLA composites were prepared in accordance with American Society for Testing and Materials D638 standards [15], as shown in Figure 4.2. A Taguchi L_9 orthogonal array experimental design was employed; the following FDM process parameters were adopted and depicted in Table 4.1.

TABLE 4.1   FDM process parameters and its levels

			LEVELS		
S. No.	PROCESS PARAMETER	SYMBOL	Level-1	Level-2	Level-3
1	Nozzle temperature in °C	NT	195	200	205
2	Layer thickness in mm	LT	0.1	0.2	0.3
3	Infill density in %	C	60	80	100
4	Filler in wt%	F	3	6	9

Abbreviation: FDM, fused deposition modelling.

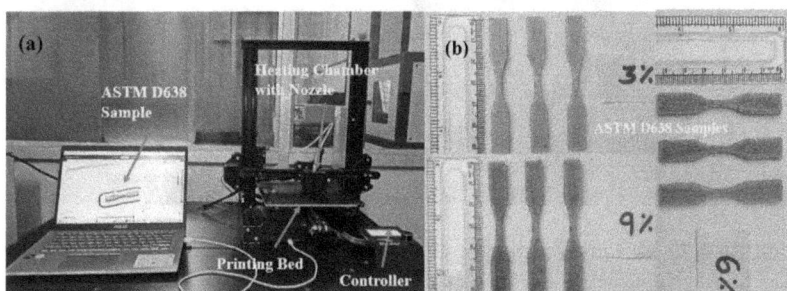

FIGURE 4.2   (a) Fused deposition modelling (FDM) setup and its components. (b) Tensile test samples printed by FDM

4.2.4 Testing Details

Tensile tests on PLA composite specimens and pure PLA specimens were conducted using an INSTRON 3367 machine, as shown in Figure 4.3, which has a load capacity of ± 20 kN and is equipped with a 40 mm axial extensometer. All tensile tests were performed at a crosshead speed of 1 mm/min, and the corresponding values of ultimate tensile strength (R_1), modulus of elasticity (R_2), and percentage of elongation (R_3) were determined for the samples. Furthermore, three samples were tested for each set of measurements. The tensile test results of PLA composite samples are depicted in Table 4.2.

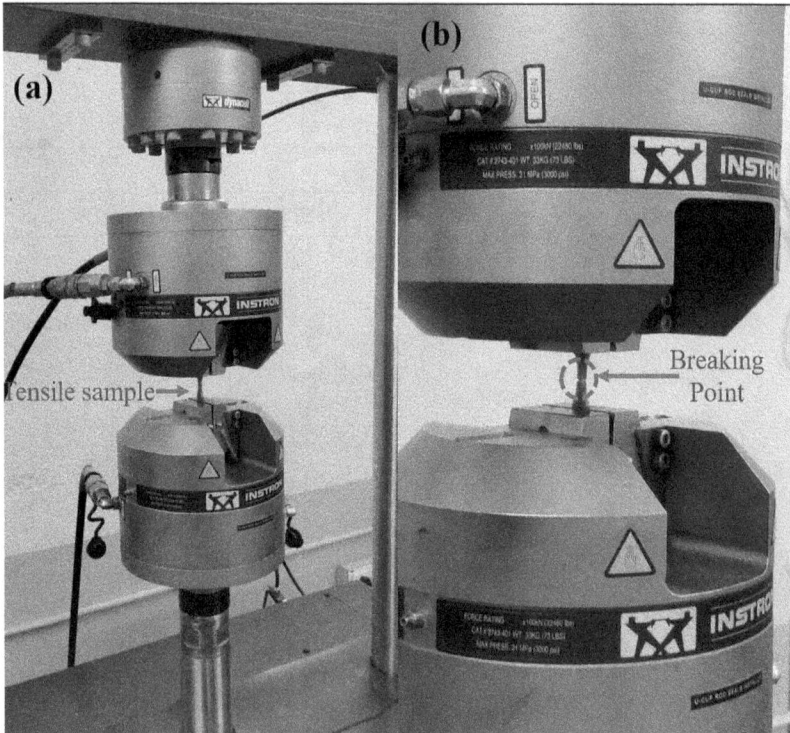

FIGURE 4.3   (a) Tensile testing of samples. (b) Breaking point of the sample

TABLE 4.2 Tensile test results of PLA-based coir fibre and additive samples printed by FDM

RUN	LT (MM)	ID (%)	NT (°C)	F(wt%)	R_1 IN MPA	R_2 IN MPA	R_3 IN (%)
1	0.1	60	195	3	62.41	717.74	5.66
2	0.1	80	200	6	70.40	827.89	6.40
3	0.1	100	205	9	74.12	293.07	1.99
4	0.2	60	200	9	71.20	720.11	5.77
5	0.2	80	205	3	55.84	827.89	6.40
6	0.2	100	195	6	56.07	293.07	1.99
7	0.3	60	205	6	60.21	746.21	5.98
8	0.3	80	195	9	63.78	595.71	6.50
9	0.3	100	200	3	57.80	575.61	4.19

Abbreviation: PLA, polylactic acid.

TABLE 4.3 ANOVA for R_1

SOURCE	DF	ADJ SS	ADJ MS	F-VALUE	P-VALUE
Regression	4	303.48	75.87	4.07	0.102
LT (mm)	1	105.337	105.337	5.64s	0.076*
ID (%)	1	5.665	5.665	0.3	0.611*
NT (°C)	1	10.428	10.428	0.56	0.496*
F (wt%)	1	182.05	182.05	9.75	0.035
Error	4	74.654	18.663		
Total	8	378.133			

Abbreviations: ANOVA, analysis of variance; DF, degrees of freedom; MS, mean square; SS, sum of squares.

4.3 RESULTS AND DISCUSSIONS

4.3.1 Statistical Analysis of R_1

Table 4.3 presents the results of a linear regression analysis of the response (R_1) to FDM process parameters. According to the findings, factors LT and F are significant at the 5% level of significance (p-value $\leq$ 0.05), and their

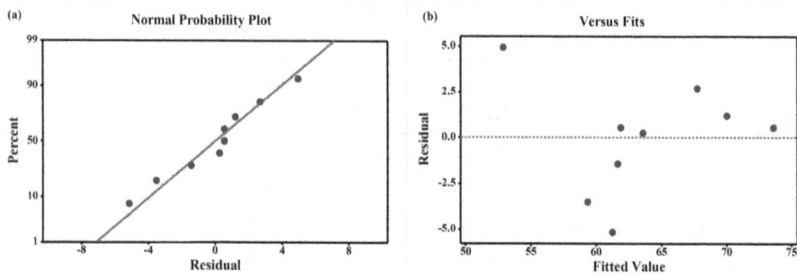

FIGURE 4.4   (a) Normal plot of residuals for R_1. (b) Residual vs. predicted plot for R_1

F-value is higher than that of the other factors [20]. The parameters indicated by (*) in Table 4.3 are the only terms or factors that are deemed statistically insignificant. Additionally, R^2 and R^2 (adj) are used to test the model's suitability, sufficiency, and fitness. The R^2 score for R_1 is 80.26%, indicating that the experimental data are good and well-fitted. Strong agreement between the experimental and predicted values is demonstrated by the modified R^2 value for R_1, which is 60.51%. Additionally, a normal probability plot is used to test the data's normality. Figure 4.4(a) shows the residuals' normal probability plot [21]. The fact that the data points are located nearer the straight line suggests that the R_1 data is regularly distributed. The relationship between residual and expected fits is displayed in Figure 4.4(b), and the data points indicate that the residuals were less structured.

Further, response obtained as per the experimental layout is used to develop the quadratic model for R_1 [21]. The actual prediction model obtained from the design expert software for R_1 is given below:

$$R_1 = 12.1 - 41.9 \, LT(mm) - 0.0486 \, ID(\%) + 0.264 \, NT(^\circ C) + 1.836 \, F \, wt.\% \quad (4.1)$$

It is evident from Eq. (4.1) that there is a positive correlation (i.e., positive effect) between R_1 and the factors NT and F. This suggests that when factors are varied from a lower range to a higher range, positive coefficients of factors result in a unit increment in R_1. A unit change in input results in a unit decrease in the value of R_1, indicating that other factors have a negative effect. Predicted vs. real plots are another way to confirm the models' correctness. Figure 4.5 shows the actual (from experiments) vs. projected (from model) figure for R_1. The plots show that the models accurately depict the correlation between the components of the factors.

Additionally, parametric analysis was done to confirm the ANOVA results (Table 4.3). The important factors that most affect the mean of R_1

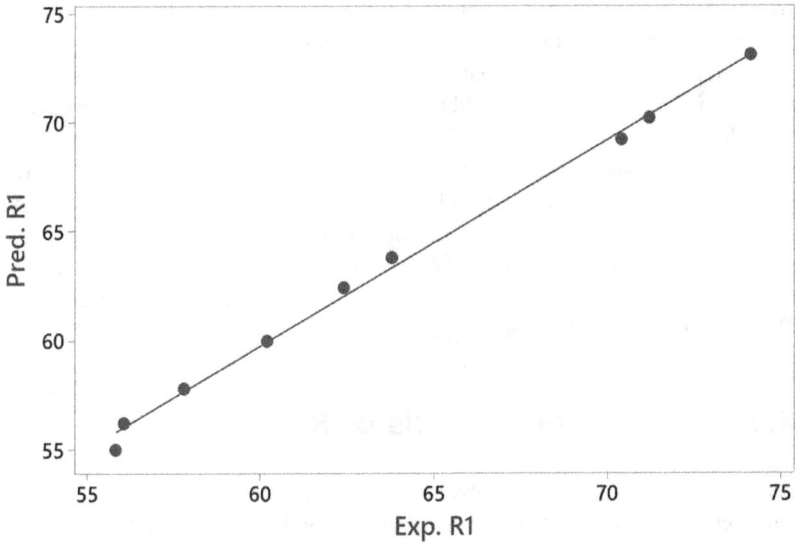

FIGURE 4.5 Predicted vs. actual plots for R_1

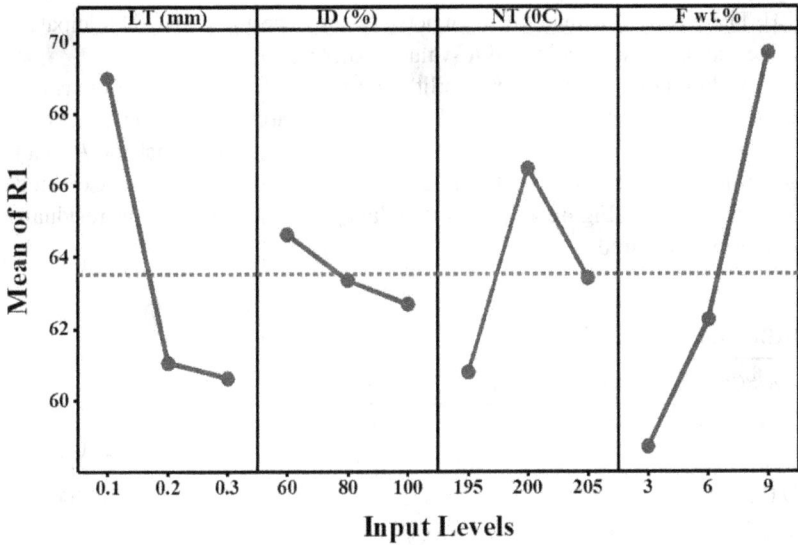

FIGURE 4.6 Main effect plot for R_1

are the parameters LT and F. Figure 4.6 showed that the value of R_1 rapidly decreases from 68.97 to 60.59 MPa when factor LT increases from 0.1 to 0.3 mm; this is due to fact that the improved interlayer adhesion between matrix and polymer with reduced residual stress accumulation [6]. However, the value of R_1 increases from 58.68 to 69.7 MPa when the factor F is changed from 3% to 9%. Here's the measure, the load-carrying capacity increases with increasing the fibre percentages [15]. This supports the LT and F ANOVA results. According to the ANOVA table and main effect map, other factors have a marginally smaller impact on the R_1. From the main effect analysis, the optimal setting for the y_1 is obtained as LT (0.1 mm, L_1), ID (60, L_1), NT (200, L_2), and F (9, L_3) to get a higher R_1 value.

4.3.2 Statistical Analysis of R_2

Linear regression is used to evaluate the response (R_2) to FDM process parameters, and the findings are given in Table 4.4. At the 5% level of significance (p-value ≤ 0.05), the results indicate that factors ID and F are significant, and their F-value is higher than that of the other factors [18]. Except for the parameters indicated by (*) in Table 4.4, all other terms and parameters are regarded as statistically unimportant. Additionally, R^2 and R^2 (adj) are used to test the model's suitability, sufficiency, and fitness. The R^2 result for R_2 is 68.09%, indicating that the experimental data are good and well fitted [21]. There is significant agreement between the experimental and anticipated values, as seen by the adjusted R^2 value of 61.25%.

Additionally, a normal probability plot is used to test the data's normality. Figure 4.7(a) shows the residual's normal probability curve. The fact that the data points are located nearer the straight line suggests that the R_2 data is regularly distributed [21]. The relationship between residual and expected fits is displayed in Figure 4.7(b), and the data points indicate that the residuals were less structured.

TABLE 4.4   ANOVA for R_2

SOURCE	DF	ADJ SS	ADJ MS	F-VALUE	P-VALUE
Regression	4	230,295	57,574	2.13	0.024
LT (mm)	1	1,036	1,036	0.04	0.854*
ID (%)	1	174,186	174,186	6.46	0.064*
NT (°C)	1	11,323	11,323	0.42	0.552*
F (wt%)	1	43,750	43,750	1.62	0.027
Error	4	107,939	26,985		
Total	8	338,234			

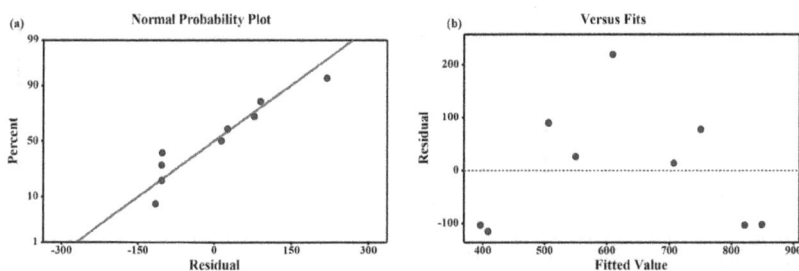

FIGURE 4.7   (a) Normal plot of residuals for R_2. (b) Residual vs. predicted plot for R_2

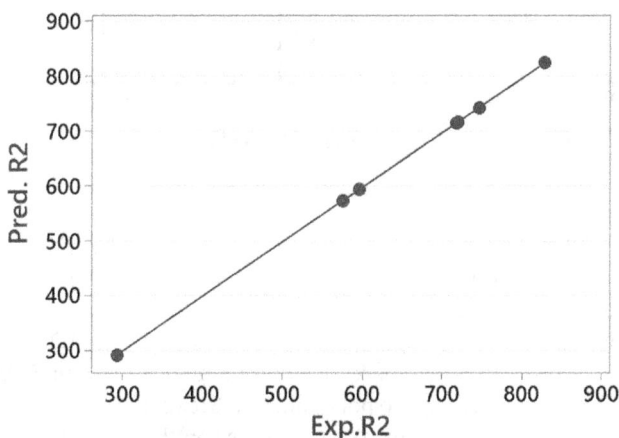

FIGURE 4.8   Predicted vs. actual plots for R_2

Further, response obtained as per the experimental layout is used to develop the quadratic model for R_2 [21]. The actual prediction model obtained from the design expert software for R_2 is given below:

$$R_2 = -290 + 131\,\mathrm{LT(mm)} - 8.52\,\mathrm{ID}(\%) + 8.7\,\mathrm{NT}(^{\circ}\mathrm{C}) - 28.5\,\mathrm{F\,wt.\%} \quad (4.2)$$

Eq. (4.2) shows that the factors LT and NT have a positive correlation (i.e., a positive effect) on R_2. This demonstrates that positive coefficients of factors cause a unit increase in responses (R_2) when these factors are changed from lower to higher ranges. Other components have a negative influence, which means that each unit change in input results in a unit decrease in the value of R_2. Further, the models' accuracy is tested using predicted vs. experimental values [19]. Figure 4.8 shows a disagreement between the actual (experimental) and expected (model-derived) plots for R_2. The graphs show that the models accurately predicted the response for the factors.

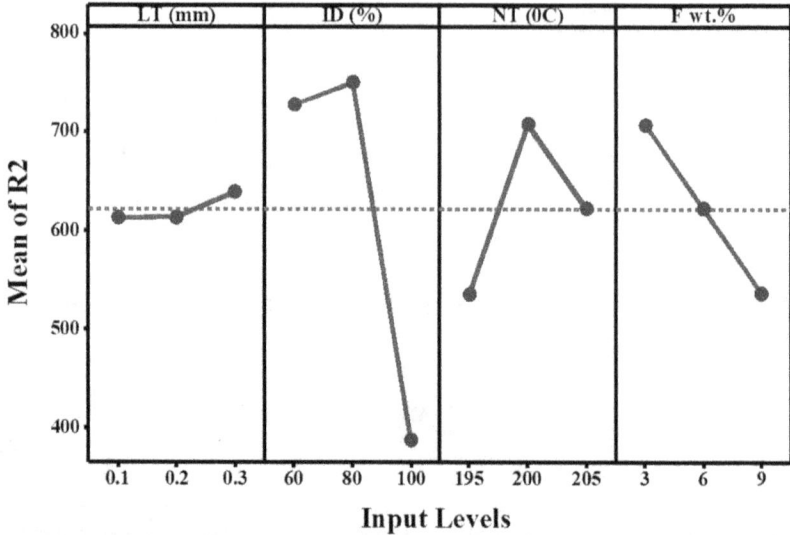

FIGURE 4.9   Main effect plot for R_2

Furthermore, parametric analysis was used to validate the ANOVA results (Table 4.4). The parameters ID and F are the most important factors influencing the mean of R_2. Figure 4.9 shows that when factor ID is increased from 60% to 100%, the value of R_2 declines dramatically, from 750.49 N to 387.25 N. This may occur due to poor adhesion between the matrix and polymer with more voids [8]. On the other hand, when the factor F is increased from 3% to 9%, the value of R_2 reduces dramatically from 707.08 N to 536.29. This is measurement, poor interfacial bonding, fibre agglomeration, and dispersion of PLA matrix continuity [13]. This supports the ANOVA results for ID and F. According to the ANOVA table and main effect plot, other factors have a minimal and less significant impact on R_2. From the main effect analysis, the optimal setting for the y_1 is obtained as LT (0.3 mm, L_3), ID (80, L_2), NT (200, L_2), and F (3, L_1) to get a higher R_2 value.

4.3.3 Statistical Analysis of R_3

The response (R_3) to FDM process parameters is assessed using linear regression, and the results are tabulated in Table 4.5. The results show that factors LT and ID are significant at a 5% and 10% level of significance (p-value $\leq$ 0.05, 0.10), and also their F-value is higher compared to the other factors. The other terms/parameters, except parameters highlighted by (*) in Table 4.1, are considered to be statistically insignificant.

TABLE 4.5 ANOVA for R_3

SOURCE	DF	ADJ SS	ADJ MS	F-VALUE	P-VALUE
Regression	4	16.0418	4.0104	1.47	0.0358
LT (mm)	1	1.1441	1.1441	0.42	0.055*
ID (%)	1	14.2296	14.2296	5.23	0.084*
NT (°C)	1	0.0081	0.0081	0	0.959
F(wt%)	1	0.66	0.66	0.24	0.648
Error	4	10.8918	2.723		
Total	8	26.9336			

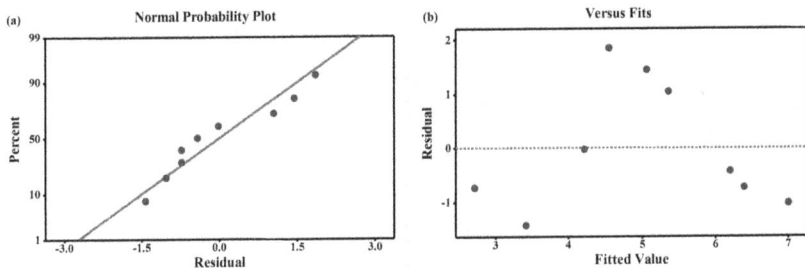

FIGURE 4.10 (a) Normal plot of residuals for R_3. (b) Residual vs. predicted plot for R_3

Additionally, the satisfactory and adequacy and fitness of the model are tested via R^2 and R^2 (adj). The R^2 value for R_3 is 85.09%, which signifies that the experimental data are well-suited and satisfactory with experimental data [19]. The adjusted R^2 value for R_3 is 80.54%, which shows the strong agreement between experimental and projected values [18]. Further, normality of the data is tested through a normal probability plot. The normal probability plot for the residual is depicted in Figure 4.10(a). The data points are found closer to the straight line, which indicates that the R_3 data is normally distributed. Figure 4.10(b) shows the relationship between residual and predicted fits and the data points signify that the residual was structure less.

Additionally, the quadratic model for R_3 is developed using the response acquired according to the experimental plan [21]. The following lists the actual prediction models that were acquired via the R_3 Minitab software:

$$R_3 = 9.5 + 4.37\,LT(mm) - 0.0770\,ID(\%) + 0.007\,NT(°C) - 0.111\,F\,wt.\% \quad (4.3)$$

Eq. (4.3) shows that the factors LT and NT have a positive correlation with R_3, meaning they have a positive influence. This shows that when factors are varied from a lower range to a higher range, positive coefficients of factors result in a unit increment in the responses (R_3) [21]. Other elements have a negative impact, meaning that a unit change in input causes a unit drop in R_3 value. Plots of projected and actual data are also used to confirm the models' correctness. There is a strong agreement between the actual and projected values for R_3 (Figure 4.11). The model correctly represents the correlation between response and factors.

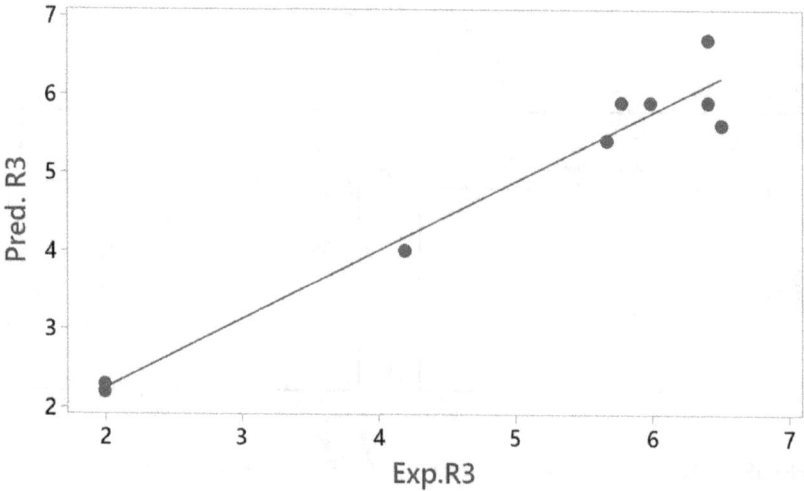

FIGURE 4.11   Predicted vs. actual plots for R_3

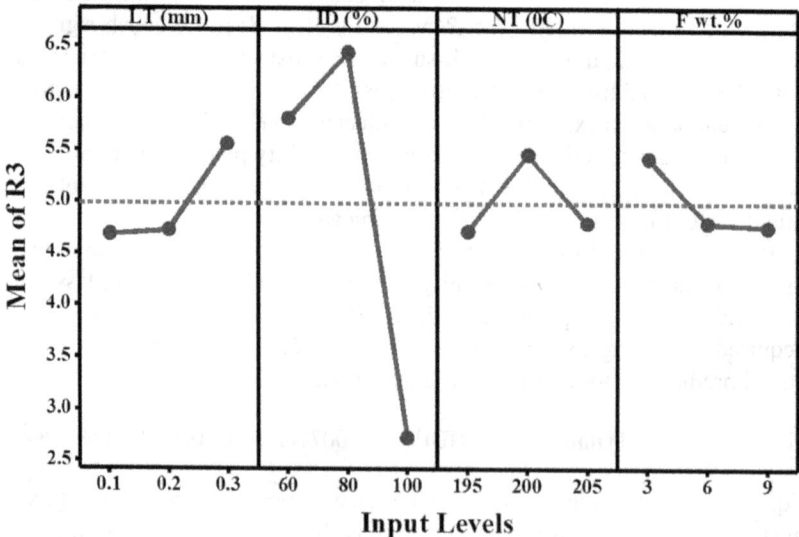

FIGURE 4.12   Main effect plot for R_3

The parametric analysis has also been done to confirm the ANOVA results (Table 4.5). The variables that most significantly affect the mean of R_3 are the parameters LT and ID. The value of R_3 increased from 4.68% to 5.55% when factor LT was increased from 0.1 to 0.3 mm, as shown in Figure 4.12. This is due to the fact that the better interfacial bonding with reduced number of interfaces and uniform distribution of load on the layers [11]. On the other hand, the value of R_3 substantially drops from 5.80% to 2.22% N when the factor ID is changed from 60% to 100%. The measure is to increase the brittleness leads to high stress concentration and crack propagation [13]. This supports the ANOVA findings for LT and ID. According to the main effect plot and ANOVA table, other factors have a minor and less significant impact on the R_3 [21]. From the main effect analysis, the optimal setting for the R_3 is obtained as LT (0.2 mm, L_2), ID (100, L_3), NT (195, L_1), and F (9, L_3) to get a higher R_3 value.

4.3.4 Result Validation

Validation is used to determine the ideal setting derived from the main effect analysis. The ideal configuration for R_1 is LT (0.1 mm, L_1), ID (60, L_1), NT (200, L_2), and F (9, L_3); for R_2, it is LT (0.3 mm, L_3), ID (80, L_2), NT (200, L_2), and F (3, L_1); and for R_3, it is LT (0.2 mm, L_2), ID (100, L_3), NT (195, L_1), and F (9, L_3). The outcomes of the re-experimentation are displayed in Table 4.6. Re-experimental values for R_1, R_2, and R_3 are found to be closer to the ideal projected values, according to the results. Less than 5% separates the experimental and model results. This indicates that the model's output is appropriate and adequate.

TABLE 4.6   Validation results of FDM process

OPTIMAL SETTING	OUTPUT PARAMETERS	REGRESSION MODEL RESULT	RE-EXP. RESULTS	% DEVIATION
LT (0.1 mm, L_1), ID (60, L_1), NT (200, L_2), and F (9, L_3)	R_1	79.14	78.03	1.4
LT (0.3 mm, L_3), ID (80, L_2), NT (200, L_2), and F (3, L_1)	R_2	938.85	915.66	2.47
LT (0.2 mm, L_2), ID (100, L_3), NT (195, L_1), and F (9, L_3)	R_3	1.95	1.88	3.68

4.4 CONCLUSION

This chapter presents the unleashing potential of FDM-printed coir-based green composites. In this case, coir fibres were used as reinforcement material and PLA serves as matrix material. The experiment used a Taguchi (L_9) orthogonal array with four FDM parameters (LT, NT, ID, and F) and three responses (R_1, R_2, and R_3). The ANOVA findings show that the factors LT, NT, ID, and F are primarily significant for R_1, R_2, and R_3 at the 5% level of significance (p-value < 0.05). This means that the main effect is significant in terms of FDM responses. Similarly, the model's satisfaction, sufficiency, and fitness are examined using R^2 and R^2 (adj), and the results show satisfactory and strong agreement between experimental and predicted values. The recommended settings for R_1 are LT (0.1 mm, L_1), ID (60, L_1), NT (200, L_2), and F (9, L_3); for R_2, it is LT (0.3 mm, L_3), ID (80, L_2), NT (200, L_2), and F (3, L_1); and for R_3, it is LT (0.2 mm, L_2), ID (100, L_3), NT (195, L_1), and F (9, L_3). Further, model diagnostic analysis is done via normal plot, residual vs. predicted plot, and seen data for R_1, R_2, and R_3; all follow a normal distribution and are independent. Additionally, model diagnostic analysis is performed via the normal plot, residual vs. predicted plot, and predicted data for R_1, R_2, and R_3; all follow a normal distribution and are independent. Finally, model validation is performed to ensure that the model is robust and that it anticipated acceptable and satisfactory results. Similarly, no significant difference between predicted and experimental results with a reduced average error is observed for the FDM process.

REFERENCES

1. Kristiawan, Ruben Bayu, Fitrian Imaduddin, Dody Ariawan, Ubaidillah, and Zainal Arifin. "A review on the fused deposition modeling (FDM) 3D printing: Filament processing, materials, and printing parameters." *Open Engineering* 11 (2021): 639–649.
2. Sathish Kumar, Adapa, and Jagadish. "Prospects of natural fiber-reinforced polymer composites for additive manufacturing applications: A review." *JOM* 75 (2023): 920–940. https://doi.org/10.1007/s11837-022-05670-w.
3. Singh, Devinder Pal. "Research in polymeric materials – Current status and future potential research in polymeric materials – Current status and future potential." *National Conference on Advances in Applied Sciences and Mathematics, NCASM 20*, Chitkara University, Rajpura, India, 24–25 Sept. 2020: 1–9.

4. Eldeeb, Ibrahim, Michael Alexandrovitch Petrov, Cezary Grabowik, Ehssan Esmael, Maher Rashad, and Salah Ebied. "Mechanical properties of PLA printed samples in different printing directions and orientations using fused filament fabrication, Part 1: Methodology." *International Conference on Intelligent Systems in Production Engineering and Maintenance*, Springer Nature Switzerland, Cham, 2023: 643–657. https://doi.org/10.1007/978-3-031-44282-7.

5. Chalgham, Ali, Andrea Ehrmann, and Inge Wickenkamp. "Mechanical properties of FDM printed PLA parts before and after thermal treatment." *Polymers (Basel)* 13, no. 8 (2021): 1239. https://doi.org/10.3390/polym13081239.

6. Shahrjerdi, Ali, Mojtaba Karamimoghadam, and Mahdi Bodaghi. "Enhancing mechanical properties of 3D-printed PLAs via optimization process and statistical modeling." *Journal of Composite Science* 7, no. 4 (2023): 1–13.

7. Patil, Shashwath, Sathish Tai, Emad Makki, and Jayant Giri. "Experimental study on mechanical properties of FDM 3D printed polylactic acid fabricated parts using response surface methodology." *AIP Advances* 14 (2024): 035125. https://doi.org/10.1063/5.0191017.

8. Kacem, Mohamed Amine, Moussa Guebailia, Mohammadreza Lalegani Dezaki, Said Abdi, Nassila Sabba, Ali Zolfagharian, and Mahdi Bodaghi. "Development and 3D printing of PLA bio-composites reinforced with short yucca fibers and enhanced thermal and dynamic mechanical performance." *Journal of Materials Research and Technology* 36 (2025): 1243–1258.

9. Wang, Kui, Yanlu Chang, Ping Cheng, Wei Wen, Yong Peng, Yanni Rao, and Said Ahzi. "Effects of PLA-type and reinforcement content on the mechanical behaviour of additively manufactured continuous ramie fiber-filled biocomposites." *Sustainability* 16, no. 7 (2024): 2635.

10. Liu, Hao, Hui He, Xiaodong Peng, Bai Huang, and Jiaxiong Li. "Three-dimensional printing of poly (lactic acid) bio-based composites with sugarcane bagasse fiber: Effect of printing orientation on tensile performance." *Polymers for Advanced Technologies* 30 (2019): 910–922. https://doi.org/10.1002/pat.4524.

11. Xiao, Xianglian, Venkata S. Chevali, Pingan Song, Dongning He, and Hao Wang. "Polylactide/hemp hurd biocomposites as sustainable 3D printing feedstock." *Composites Science and Technology* 184 (2019): 107887. https://doi.org/10.1016/j.compscitech.2019.107887.

12. Mohammadsalih, Zaid G., Muhammad Muawwidzah, Vasi Uddin Siddiqui, and S. M. Sapuan. "Mechanical properties of wood fibre filled polylactic acid (PLA) composites using additive manufacturing techniques." *Journal of Natural Fiber Polymer Composites* 3 (2023): 2821–3289.

13. Depuydt, Delphine, Michiel Balthazar, Kevin Hendrickx, Wim Six, Eleonora Ferraris, Frederik Desplentere, Jan Ivens, and Aart W. Van Vuure. "Production and characterization of bamboo and flax fiber reinforced polylactic acid filaments for fused deposition modeling (FDM)." *Polymer Composites* 40 (2019): 1951–1963. https://doi.org/10.1002/pc.24971.

14. Tsou, Chi-Hui, Wei-Hua Yao, Chin-San Wu, Chih-Yuan Tsou, Wei-Song Hung, Jui-Chin Chen, Jipeng Guo, Shuai Yuan, Ehua Wen, Ruo-Yao Wang, Maw-Cheng Sunn, Shi-Chih Liu, and Manuel Reyes De Guzman. "Preparation and characterization of renewable composites from polylactide and rice husk for 3D printing applications." *Journal of Polymer Research* 26 (2019): 227.

15. Patil, Milind, Mugdha Dongre, D. N. Raut, and Ajinkya Naik. "Multi-objective optimization of fused filament fabrication (FFF) parameters for rice husk reinforced PLA composites." *Next Materials* 8 (2025): 100540. https://doi.org/10.1016/j.nxmate.2025.100540.

16. Muthu Natarajan, S. S. Senthil, and Pandiarajan Narayanasamy. "Investigation of mechanical properties of FDM-processed acacia concinna–filled polylactic acid filament." *International Journal of Polymer Science* 2022, no. 1 (2022): 4761481. https://doi.org/10.1155/2022/4761481.

17. Jang, Hyunh, Sangwoo Kwon, Sun Jong Kim, and Su-il Park. "Maleic anhydride-grafted PLA preparation and characteristics of compatibilized PLA/PBSeT blend films." *International Journal of Molecular Sciences* 23, no. 13 (2022): 7166.

18. Adapa, Sathish Kumar, Jagadish, and Srinivasu Gangi Setti. "Optimization of twin screw melt mixer setup pro cess parameters for better blending of polymers and polymer composites for FDM applications." *Next Materials* 8 (2025): 100671. https://doi.org/10.1016/j.nxmate.2025.100671.

19. Singh, Ashangbam Satyavrata, Sudipta Halder, Jialai Wang, and Jagadish. "Extraction of bamboo micron fibres by optimized mechano-chemical process using a central composite design and their surface modification." *Materials Chemistry and Physics* 199 (2017): 23–33. https://doi.org/10.1016/j.matchemphys.2017.06.040.

20. Mirza, Faizaan, Satish Baloor Shenoy, Srinivas Nunna, Chandrakant Ramanath Kini, and Claudia Creighton. "Effect of material extrusion process parameters on tensile performance of pristine and discontinuous fibre reinforced PLA composites: A review." *Progress in Additive Manufacturing* 10 (2025): 3251–3265. https://doi.org/10.1007/s40964-024-00825-4.

21. Jagadish, Maran Rajakumaran, and Amitava Ray. "Investigation on mechanical properties of pineapple leaf–based short fiber–reinforced polymer composite from selected Indian (northeastern part) cultivars." *Journal of Thermoplastic Composite Materials* 33 (2020): 324–342. https://doi.org/10.1177/0892705718805535.

Advancing Sustainability with Statistical Modelling of FDM-Printed Cane-Based Green Composites

5

5.1 INTRODUCTION

In recent years, additive manufacturing (AM) has emerged as a transformative technology for producing complex geometries without the tooling. This method provides several benefits over traditional subtractive manufacturing methods, including zero wastage of raw material, reduced fabrication time, and minimal energy consumption [1]. These advantages make AM highly appropriate for meeting the needs of various industries, small-scale vendors, and researches alike. Also, AM encloses a range of techniques for fabricating objects using materials such as metals, ceramics, and polymers. Among these,

DOI: 10.1201/9781003732143-5

polymers are widely used in small-scale applications due to their enormous properties like light in weight, cost-effective, adequately available, and often require minimal post-processing [2]. These characteristics make polymers an ideal choice for various applications, especially where efficiency and affordability are prioritized.

Despite their advantages, polymers have certain issues like being toxic in nature, non-biodegradable, and exhibiting poor mechanical and thermal properties. To overcome these limitations, polymers are commonly reinforced with natural fibres, which helps enhance their properties as compared to virgin polymers [2]. Additionally, natural fibres are being relatively low cost, abundantly available, easy to process, fast to cure, biodegradable, and possess superior mechanical properties. Wang et al. [3] investigated the mechanical properties of polylactic acid (PLA)–based composites reinforced with ramie yarn fibres, using samples printed by the fused deposition modelling (FDM) process. The results showed that the tensile strength and Young's modulus of the PLA composite biopolymer samples increased by 76.65% and 160.84%, respectively, compared to virgin PLA samples printed using the same method.

Kacem et al. [4] studied the mechanical properties of PLA samples reinforced with yucca fibres, fabricated using the FDM process. The results showed that the tensile strength and modulus increased to 61.06 and 1,316 MPa, respectively, compared to the properties of pure PLA samples, which were 46.42 and 1,303 MPa. Siddiqui et al. [5] investigated the mechanical and flammability properties of wood-based PLA samples fabricated using the FDM process. The results revealed that the bio-composite samples exhibited lower mechanical properties and higher flammability compared to virgin PLA samples printed by FDM.

Moreover, limited research has been conducted on optimizing the FDM process parameters for PLA-based polymers reinforced with natural fibres and additives to improve the mechanical properties of samples printed using FDM [6].

Patil et al. [7] investigated the optimization of FDM process parameters for PLA reinforced with rice husk powder filament used as a feedstock material. The results revealed that the mechanical properties were enhanced under optimal FDM parameters: a nozzle temperature of 190°C, a layer thickness (LT) of 0.2 mm, and a rice husk content of 1 wt%. Natarajan et al. [8] investigated the mechanical properties of PLA-based filaments reinforced with *Acacia concinna* (AC) for FDM applications. The results showed that the tensile strength improved in PLA samples filled with 25 wt% AC compared to virgin PLA samples. The optimal process parameters for enhanced mechanical properties were found to be an LT of 0.16 mm, an infill density (ID) of 100%, and a print speed of 50 mm/s.

Limited research has been conducted on optimizing FDM process parameters using PLA polymer reinforced with natural fibres and additives as feedstock material [9]. This chapter explores the advancement of sustainability through statistical modelling of FDM-printed cane bamboo green composites. Initially, cane bamboo stems were processed to obtain cane bamboo powder, which was then treated with a sodium hydroxide (NaOH) solution to remove unwanted chemical constituents and clean the fibre surfaces. Subsequently, a PLA-based composite filament reinforced with cane bamboo powder and additives was prepared using an internal melt mixer and a filament extruder setup. Further, tensile test samples of the PLA composite were then fabricated using the FDM process. Also, mechanical characterization was performed on these PLA-based composite samples and compared with pure PLA samples produced under the same conditions. Finally, statistical analysis was conducted to evaluate the influence of FDM process parameters on the mechanical properties of the printed samples.

5.2 MATERIALS AND METHODS

5.2.1 Thermo-Polymers and Coupling Agents

Bio-PLA with the code KBB1012 was used as the matrix polymer material. Short bamboo cane fibres (approximately 300 μm in size) were selected as the reinforcement material at different weight percentages, i.e., 3%, 6%, and 9% respectively. To ensure proper adhesion between the matrix and the reinforcement, coupling agents such as malic anhydride (2 wt%, density 1.48 g/cm³) and polyethylene glycol (6 wt%, density 1.125 g/cm³) were adopted [10]. Also, NaOH solution, acetone, and deionized water are used for alkali treatment on cane bamboo fibre's surface treatment [11].

5.2.2 Processing of Bamboo Cane Fibres

Bamboo cane plants exported from the Bakali Hills, Assam, India, are initially processed by cutting them into small stakes. These stakes are then subjected to shearing operations to produce raw bamboo cane powder. To ensure uniform particle size, the powder is sieved using a 300 μm mesh [11]. The remaining coarse particles are further ground until all the powder passes

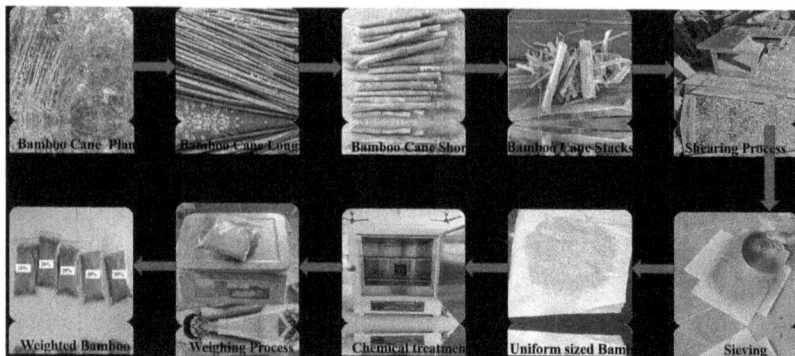

FIGURE 5.1 Sequence of operations for processing of bamboo cane fibres

through the mesh, achieving a consistent particle size. An alkali treatment is then performed on the bamboo cane fibres using a 5 wt% NaOH solution, with stirring at 80 rpm at 50°C [12]. After the treatment, the fibres are washed with acetone to remove any unreacted NaOH, followed by thorough rinsing with deionized water until the pH drops below 7. The resulting wet powder is dried in a hot air oven at 60°C for 24 hours. Finally, the dried powder is sealed in airtight packaging for future use. The bamboo cane fibre processing steps are shown in Figure 5.1.

5.2.3 Preparation of Samples

The PLA-based composite filament, reinforced with bamboo cane fibres and additives, is produced using a multi-screw filament extruder setup. This composite filament serves as the feedstock material for FDM. An Ender 3 FDM printer is used to fabricate both the PLA-based bamboo cane composite samples and pure PLA samples, following the American Society for Testing and Materials D638 standard specifications [13]. The samples were printed (as shown in Figure 5.2) based on a Taguchi L_9 orthogonal array design, and the corresponding FDM process parameters and their levels are presented in Table 5.1.

5.2.4 Testing Details

Tensile tests on PLA composite specimens and pure PLA specimens were conducted using an INSTRON 3367 machine, as shown in Figure 5.3, which has a load capacity of ±20 KN and is equipped with a 40 mm axial

TABLE 5.1   FDM process parameters and their levels

S. NO.	PROCESS PARAMETER	SYMBOL	LEVELS		
			LEVEL-1	LEVEL-2	LEVEL-3
1.	Filler in wt%	x_1	3	6	9
2.	Thickness in mm	x_2	0.1	0.2	0.3
3.	Nozzle Temperature in °C	x_3	195	200	205
4.	Infill Density in %	x_4	60	80	100

Abbreviation: FDM, fused deposition modelling.

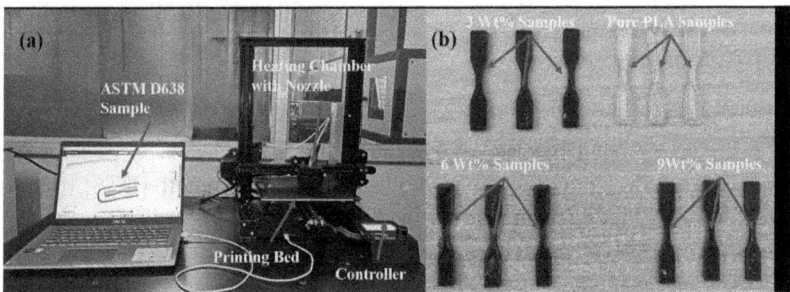

FIGURE 5.2   (a) Fused deposition modelling (FDM) printer and its components. (b) Polylactic acid (PLA) and PLA composite samples printed by FDM

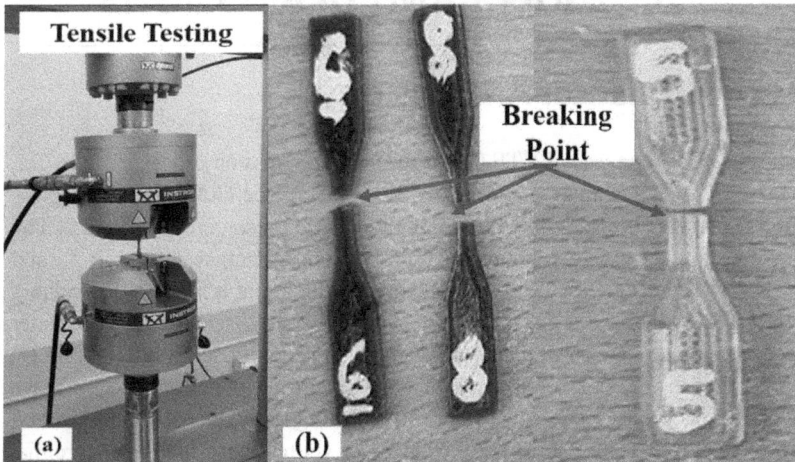

FIGURE 5.3   (a) Tensile testing setup. (b) Fractured tensile specimens

TABLE 5.2   Tensile test results of PLA-reinforced cane bamboo and additive composite specimens

EX. NO.	x_1	x_2	x_3	x_4	R_2:Y_1	R_2:Y_2
1	3	0.1	195	60	69.86	847.48
2	3	0.2	200	80	71.79	801.40
3	3	0.3	205	100	63.74	900.41
4	6	0.1	200	100	71.28	947.61
5	6	0.2	205	60	74.67	721.66
6	6	0.3	195	80	71.78	665.78
7	9	0.1	205	80	45.46	737.23
8	9	0.2	195	100	67.74	627.59
9	9	0.3	200	60	76.65	677.47

Abbreviations: PLA, polylactic acid.

extensometer. All tensile tests were performed at a crosshead speed of 1 mm/min, and the corresponding values of ultimate tensile strength (UTS; y_1) and modulus of elasticity (y_2) are determined for the samples. Furthermore, three samples were tested for each set of measurements. The tensile test results of PLA composite samples are depicted in Table 5.2.

5.3 RESULTS AND DISCUSSIONS

5.3.1 Statistical Analysis of y_1

The statistical analysis of UTS (y_1) of the FDM process on process factors, i.e., x_1, x_2, x_3, and x_4 is analysed using analysis of variance (ANOVA) and regression analysis [7]. The ANOVA findings for y_1 are presented in Table 5.3. The results show that the components x_1, x_2, x_3 (linear effect), $x_1 \times x_2$, $x_1 \times x_3$, and $x_2 \times x_3$ (interaction effect) are statistically significant at a 5% level (p-value $\leq$ 0.05) [6]. Similarly, the F-values for these factors are discovered to be greater and statistically significant. The other factors except those highlighted by (*) in Table 5.3 are considered to be statistically insignificant and do not have much impact on y_1. The coefficient of determination value (R^2) and adjusted coefficient of determination value R^2 (adj) for y_1 are determined to ensure that the model is satisfactory, adequate, and fit for purpose. The R^2 score for y_1 is 93.28%, indicating that the experimental data are well fitted and satisfactory to

TABLE 5.3 ANOVA for y_1

SOURCE	DF	ADJ SS	ADJ MS	F-VALUE	P-VALUE
Regression	7	1,407.7	201.1	19.83	0*
x_1	1	153.11	153.108	15.1	0.003*
x_2	1	156.54	156.542	15.44	0.003*
x_3	1	283.26	283.26	27.93	0*
x_4	1	0.18	0.179	0.02	0.897
$x_1 \times x_2$	1	389.25	389.252	38.39	0*
$x_1 \times x_3$	1	185.54	185.541	18.3	0.002*
$x_2 \times x_3$	1	181.73	181.727	17.92	0.002*
Error	10	101.4	10.14		
Lack-of-fit	1	42.63	42.626	6.53	0.031
Pure error	9	58.78	6.531		
Total	17	1,509.1			

$R^2 = 93.28\%$, R^2-adj = 88.58%
Abbreviations: ANOVA, analysis of variance; DF, degrees of freedom; MS, mean square; SS, sum of squares.

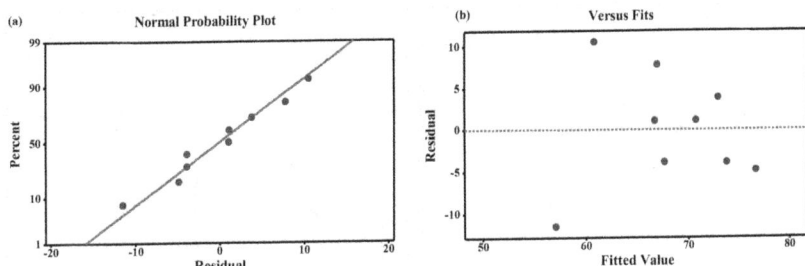

FIGURE 5.4 (a) Normal plot of residuals for y_1. (b) Residual vs. predicted plot for y_1

the experimental data [14]. The modified R^2 value for y_1 is 88.58%, indicating a significant agreement between the experimental and anticipated values.

Furthermore, the model diagnostic analysis is used to graphically analyse the response (y_1) with a normal plot and a residual vs. predicted plot. The normal probability plot for the residual is shown in Figure 5.4(a), and the data points are closer to the straight line, indicating that the y_1 data follows a normal distribution. Figure 5.4(b) illustrates the link between residual and anticipated fits. The graphical representation demonstrates that data points indicate that the residuals were structureless and independent [15].

Additionally, factor analysis is used to validate the ANOVA results. According to the factor analysis, the most significant factors influencing y_1 are x_1, x_2, and x_3. As shown in Figure 5.5, when factor x_1 varies from 3% to 6%, the value of y_1 increases drastically from 70.256 to 74.486 MPa, followed by a sharp decrease from 74.486 to 64.984 MPa when level 6–9%. This is related to the fact that the increase in tensile strength from 3 to 6 wt% fibre content is likely due to improved load transfer between the fibres and the matrix material, leading to better stress distribution. The decrease in strength beyond 6 wt% suggests that higher fibre content can lead to issues such as poor fibre-matrix bonding, fibre agglomeration, or increased porosity, which can act as stress concentrators and reduce overall strength [16]. At larger values of x_2, the factor x_2 has a considerable effect on y_1. The starting value of y_1 is 63.836 MPa at 0.1 mm, but it rapidly increases to 73.21 MPa at 0.2 and 0.3 mm. This may be the reason for improved interlayer adhesion, less residual stress accumulation, and more material continuity [17]. Similarly, the influence of x_3 on y_1 appears to be stronger (75.16 MPa) at 200°C before dropping dramatically to 62.00 MPa at 205°C. This could be the cause of overheating and degradation of polymers and its composites, increased flowability, reduced interlayer pressure, and more thermal warping [8]. On the other hand, the impact of x_4 on y_1 appears to be negative and marginally significant. The ideal configuration for y_1 is x_1 (6%, L_2), x_2 (0.2 mm, L_2), x_3 (200, L_2), and x_4 (60, L_1), according to the parametric analysis.

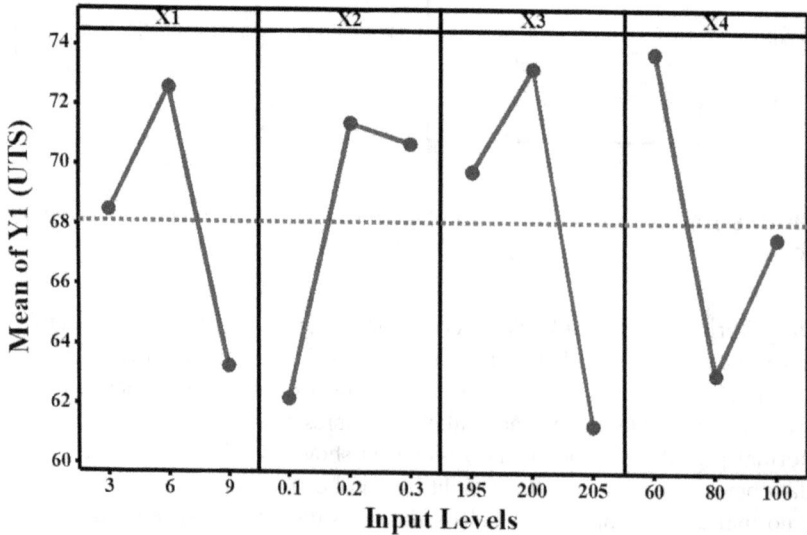

FIGURE 5.5 Main effect for y_1

At last, a quadratic model for y_1 based on the experimental layout is developed [15]. The prediction model obtained from the Minitab software for y_1 is given below:

$$y_1 = -1,743 + 153.6x_1 + 3,145x_2 + 9.53x_3 - 0.016x_4 + 60.89x_1 \times x_2 \quad (5.1)$$
$$- 0.841\ x_1 \times x_3 - 17.65\ x_2 \times x_3$$

According to Eq. (5.1), the components x_1, x_2, and x_3 have a positive correlation (i.e., positive effect) with y_1, but x_4 has a negative correlation with y_1. The positive correlation shows that a unit increment in factors x_1, x_2, and x_3 results in a unit increment in y_1. In the instance of negative correlation, a unit increase in component x_4 results in a unit decrease in factor y_1. Furthermore, the interaction impact of $x_1 \times x_2$ is positive, whereas the interactions $x_1 \times x_3$ and $x_2 \times x_3$ are negatively associated [18].

5.3.2 Statistical Analysis of y_2

The statistical analysis for y_2 is done through ANOVA and regression analysis. The ANOVA results of y_2 are tabulated in Table 5.4. The ANOVA results show that x_1, x_2, x_3, and x_4 (linear effect), $x_1 \times x_2$, $x_1 \times x_3$, and $x_2 \times x_3$ (interaction effect) are found statistically significant at a 5% level of significance (p-value ≤ 0.05) [11]. The other factors except those highlighted by (*) in Table 5.4 are considered to be statistically insignificant and do not have much

TABLE 5.4 ANOVA for y_2

SOURCE	DF	ADJ SS	ADJ MS	F-VALUE	P-VALUE
Regression	7	195,672	27,953.1	41.02	0
x_1	1	22,761	22 761.2	33.4	0*
x_2	1	3,558	3,558	5.22	0.045*
x_3	1	5,683	5,683.1	8.34	0.016*
x_4	1	54,241	54,241.2	79.59	0*
$x_1 \times x_2$	1	25,311	25,310.8	37.14	0*
$x_1 \times x_3$	1	20,129	20,129.3	29.54	0*
$x_2 \times x_3$	1	2,178	2,177.9	3.2	0.104
Error	10	6,815	681.5		
Lack-of-fit	1	5,684	5,684	45.24	0
Pure error	9	1,131	125.6		
Total	17	202,486			

$R^2 = 96.63\%$, R^2(adj)= 94.28%

impact on y_2. Next, satisfactory and adequacy as well as fitness of the model for y_2 are checked via R^2 and R^2 (adj) values. The R^2 value for y_2 is obtained as 96.63%, which signifies that the experimental data are well suited and satisfactory with experimental data [19]. The adjusted R^2 value for y_2 is obtained as 94.28%, which shows the strong agreement between experimental and expected values.

Furthermore, model diagnostic analysis is performed to graphically examine the response (y_2) model using a normal plot and a residual vs. predicted plot [15]. The normal probability plot for residuals is shown in Figure 5.6(a), and data points are closer to the straight line, indicating that the y_1 data follows a normal distribution. Figure 5.6(b) illustrates the link between residual and expected fits. The graphical representation demonstrates that data points indicate that the residuals were structureless and independent.

In addition, factor analysis was used to corroborate the ANOVA results. The factor analysis shows that x_1, x_2, x_3, and x_4 have the largest impact on y_2. Figure 5.7 shows that factor x_1 has a higher effect on y_2. Because the rate of y_2 was first greater (858.53 MPa) at 3% and subsequently significantly fell to 687.71 MPa at 9%. This is because poor interfacial bonding between the fibres and matrix results in reduced tensile modulus [20]. Factor x_2 behaves similarly to x_1. When x_2 varies from 0.1 to 0.2 mm, the rate of y_2 decreases from 852.72 to 724.19 MPa. However, as the range of x_2 increases from 0.2 to 0.3 mm, a slight increase in y_2 is observed. This could be due to poor interlayer diffusion with increased voids and reduced print pressure, which resulted in reduced tensile modulus [21]. Similarly, when x_3 is varied from 195 to 200, the effect on y_1 is positive from 720.90 to 817.08 MPa, followed by a moderate decrement of 817.08 to 794.46 MPa when x_3 is varied from 200 to 205. This could be the allowing of better wetting and penetration and reduced voids, which resulted in improved tensile modulus. However, the thermal degradation of

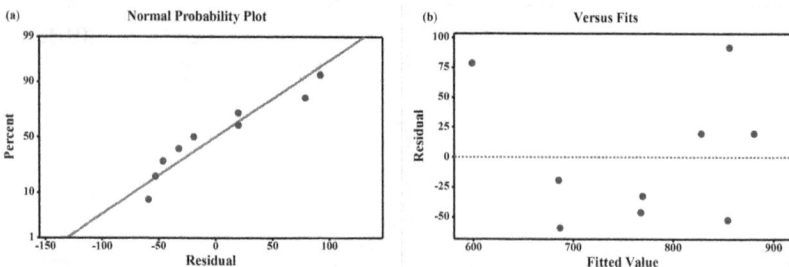

FIGURE 5.6   (a) Normal plot of residuals for y_2. (b) Residual vs. predicted plot for y_2

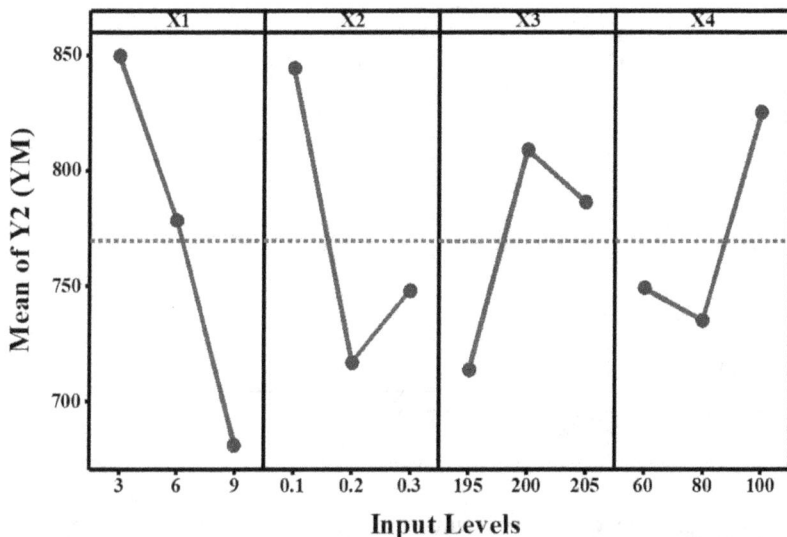

FIGURE 5.7   Main effect plot for y_2

natural fibres and polymers with excessive flow leads to reduced structural integrity [6]. On the other hand, x_4 appears to have a significant effect on y_2. When x_4 is changed from 60 to 80, the value of y_2 decreases somewhat, and when x_4 is changed from 80 to 100, y_2 increases suddenly. This could be the cause of increased structural discontinuities with irregular fibre distribution that resulted in decreased tensile modulus. However, the tensile modulus is improved at higher IDs due to improved material continuity and reduced void formation between the polymer and matrix [7]. From the factor analysis, the optimal setting for the y_2 is obtained as x_1 (3%, L_1), x_2 (0.1 mm, L_1), x_3 (200, L_2), and x_4 (100, L_3) to get the optimal y_2 value.

At last, a quadratic model for y_2 based on the experimental layout is developed [15]. The prediction model obtained from the Minitab software for y_1 is given below:

$$y_1 = 9,296 - 1,873x_1 - 14,996x_2 - 42.7x_3 + 8.894x_4 + 491.0x_1 \times x_2 \quad (5.2)$$
$$+ 8.76x_1 \times x_3 + 61.1x_2 \times x_3$$

From Eq. (5.2), it is seen that the factors, x_1, x_2, and x_3 have a negative correlation (i.e., negative effect) with y_2, and x_4 has a positive correlation with y_2. The positive correlation indicates that unit increment in factor x_4 leads to a unit increment in y_2 positively [17]. On the other hand, a unit increment of

TABLE 5.5   Model validation results

OPTIMAL SETTING	OUTPUT PARAMETERS	REGRESSION MODEL RESULT	RE-EXP. RESULTS	ARD%
x_1 (6%, L_2), x_2 (0.2 mm, L_2), x_3 (200, L_2), and x_4 (60, L_1)	y_1	76.65	76.02	0.98
x_1 (3%, L_1), x_2 (0.1 mm, L_1), x_3 (200, L_2), and x_4 (100, L_3)	y_2	729.02	728.23	0.07

Abbreviation: ARD, absolute relative deviation.

factors x_1, x_2, and x_3 leads to a unit decrement of y_2 in the case of negative correlation. Further, the interaction effect by $x_1 \times x_2$, $x_1 \times x_3$, and $x_2 \times x_3$ has a positive effect on y_2.

5.3.3 Model Validation

Model validation is used to determine the robustness of the regression model. The model is validated through re-experimentation and prediction of results using Eqs. (5.1) and (5.2). The ideal configuration determined by factor analysis for y_1 and y_2 is used for re-experimentation. Table 5.5 shows the predicted and re-experimented findings [7]. To assess model predictability, the absolute relative deviation values for y_1 and y_2 are calculated. The results indicate that re-experimental values for y_1 and y_2 are closer to the ideal projected values. The experimental and model results are relatively close, with a variation of less than 5%. This indicates that the model's results are satisfactory and acceptable [6,17]:

$$\text{ARD } (\%) = \frac{\text{Observed Value} - \text{Fitted Value}}{\text{Observed Value}} \times 100 \tag{5.3}$$

5.4 CONCLUSIONS

This chapter presented the development and statistical analysis of FDM process parameters on printing of hybrid green composites. In this case, bamboo cane is used as a reinforcement material and PLA is chosen as matrix material. The experiment used a Taguchi (L_9) orthogonal array with four FDM parameters (x_1, x_2, x_3, and x_4) and two responses (y_1 and y_2). The ANOVA findings show that the factors x_1, x_2, and x_3 are primarily significant for y_1 and y_2 at the 5% level of significance (p-value < 0.05). This means that the main effect is significant in terms of FDM responses. Similarly, the model's satisfaction, sufficiency, and fitness are examined using R^2 and R^2 (adj), and the results show satisfactory and strong agreement between experimental and predicted values. The recommended settings for y_1 are x_1 (6%, L_2), x_2 (0.2 mm, L_2), x_3 (200, L_2), and x_4 (60, L_1), whereas for y_2 they are x_1 (3%, L_1), x_2 (0.1 mm, L_1), x_3 (200, L_2), and x_4 (100, L_3). Further, model diagnostic analysis is done via normal plot and residual vs. predicted plot, and the seen data for y_1 and y_2 follow a normal distribution and are independent. Additionally, model diagnostic analysis is performed via the normal plot, residual vs. predicted plot, and predicted data for y_1 and y_2; all follow a normal distribution and are independent. Finally, model validation is performed to ensure that the model is robust and that it anticipated acceptable and satisfactory results. Similarly, no significant difference between predicted and experimental results with a reduced average error is observed for the FDM process.

REFERENCES

1. Dubey, Dushyant, Satinder Paul Singh, and Bijoya Kumar Behera. "A review on recent advancements in additive manufacturing techniques." *Proceedings of the Institution of Mechanical Engineers, Part E: Journal of Process Mechanical Engineering* (2024): 1–23. https://doi.org/10.1177/09544089241275860

2. Adapa, Sathish Kumar, and Jagadish. "Prospects of natural fiber-reinforced polymer composites for additive manufacturing applications: A review." *JOM* 75 (2023): 920–940. https://doi.org/10.1007/s11837-022-05670-w.

3. Wang, Kui, Yanlu Chang, Ping Cheng, Wei Wen, Yong Peng, Yanni Rao, and Said Ahzi. "Effects of PLA-type and reinforcement content on the mechanical behavior of additively manufactured continuous ramie fiber-filled biocomposites." *Sustainability* 16, no. 7 (2024): 2635.

4. Kacem, Mohamed Amine, Moussa Guebailia, Mohammadreza Lalegani Dezaki, Said Abdi, Nassila Sabba, Ali Zolfagharian, and Mahdi Bodaghi. "Development and 3D printing of PLA bio-composites reinforced with short yucca fibers and enhanced thermal and dynamic mechanical performance." *Journal of Materials Research and Technology* 36 (2025): 1243–1258.

5. Siddiqui, Vasi Uddin, J. Yusufa, S. M. Sapuana, Mohammad Zaid Hasana, Muhammad Muawwidzah Mudah Bistaria, and Zaid G. Mohammad Salih. "Mechanical properties and flammability analysis of wood fiber filled polylactic acid (PLA) composites using additive manufacturing." *Journal of Natural Fibers* 21 (2024): 2409868. https://doi.org/10.1080/15440478.2024.2409868.

6. Mirza, Faizaan, Satish Baloor Shenoy, Srinivas Nunna, Chandrakant Ramanath Kini, and Claudia Creighton. "Effect of material extrusion process parameters on tensile performance of pristine and discontinuous fibre reinforced PLA composites: A review." *Progress in Additive Manufacturing* 10 (2025): 3251–3265. https://doi.org/10.1007/s40964-024-00825-4.

7. Patil, Milind, Mugdha Dongre, D. N. Raut, and Ajinkya Naik. "Multi-objective optimization of fused filament fabrication (FFF) parameters for rice husk reinforced PLA composites." *Next Materials* 8 (2025): 100540. https://doi.org/10.1016/j.nxmate.2025.100540.

8. Muthu Natarajan, S., S. Senthil, and Pandiarajan Narayanasamy. "Investigation of mechanical properties of FDM-processed acacia concinna–filled polylactic acid filament." *International Journal of Polymer Science* 2022, no. 1 (2022): 4761481. https://doi.org/10.1155/2022/4761481.

9. Deb, Disha, and J. M. Jafferson. "Natural fibers reinforced FDM 3D printing filaments." *Materials Today: Proceedings* 46 (2021): 1308–1318. https://doi.org/10.1016/j.matpr.2021.02.397.

10. Nonkrathok, Wasana, Tatiya Trongsatitkul, and Nitinat Suppa Karn. "Role of maleic anhydride grafted poly (lactic acid) in improving shape memory properties of their more pensive poly(ethylene glycol) and *poly(lactic acid)* blends." Polymers (Basel) 14 (2022): 3923. https://doi.org/10.3390/polym14183923.

11. Adapa, Sathish Kumar, Jagadish, and Srinivasu Gangi Setti. "Optimization of twin screw melt mixer setup process parameters for better blending of polymers and polymer composites for FDM applications." *Next Materials* 8 (2025): 100671. https://doi.org/10.1016/j.nxmate.2025.100671.

12. Singh, Ashangbam Satyavrata, Sudipta Halder, Jialai Wang, and Jagadish. "Extraction of bamboo micron fibers by optimized mechano-chemical process using a central composite design and their surface modification." *Materials Chemistry and Physics* 199 (2017): 23–33. https://doi.org/10.1016/j.matchemphys.2017.06.040.

13. Liu, Hao, Hui He, Xiaodong Peng, Bai Huang, and Jiaxiong Li. "Three-dimensional printing of poly (lactic acid) bio-based composites with sugarcane bagasse fiber: Effect of printing orientation on tensile performance." *Polymers for Advanced Technologies* 30 (2019): 910–922. https://doi.org/10.1002/pat.4524.

14. Kopar, Mehmet, and Ali Riza Yildiz. "Experimental investigation of mechanical properties of PLA, ABS, and PETG 3-d printing materials using fused deposition modeling technique." *Materials Testing* 65, no. 12 (2023): 1–10. https://doi.org/10.1515/mt-2023-0202.

15. Jagadish, Maran Rajakumaran, and Amitava Ray. "Investigation on mechanical properties of pineapple leaf–based short fiber–reinforced polymer composite from selected Indian (northeastern part) cultivars." *Journal of Thermoplastic Composite Materials* 33 (2020): 324–342. https://doi.org/10.1177/0892705718805535.

16. Ahmad, Noesanto Dewantoro, Kusmono, Muhammad Waziz Wildan, and Herianto. "Preparation and properties of cellulose nanocrystals-reinforced poly (lactic acid) composite filaments for 3D printing applications." *Results in Engineering* 17 (2023): 100842. https://doi.org/10.1016/j.rineng.2022.100842.

17. Mohamed, Omar Ahmed, Syed Hasan Masood, and Jahar Lal Bhowmik. "Optimization of fused deposition modeling process parameters: A review of current research and future prospects." *Advances in Manufacturing* 3 (2015): 42–53. https://doi.org/10.1007/s40436-014-0097-7.

18. Mahović Poljaček, Sanja, Davor Donevski, Tamara Tomašegoviš, Urška Vrabiš Brodnjak, and Mirjam Leskovšek. "Mechanical properties of 3D printed PLA structures observed in framework of different rotational symmetry orders in infill patterns." *Symmetry* 17, no. 3 (2025): 466.

19. Gauss, Christian, Kim L. Pickering, Nina Graupner, and Jörg Müssig. "3D printed polylactide composites reinforced with short lyocell fibres–Enhanced mechanical proper ties based on bio inspired fibre fibrillation and post print annealing." *Advances in Manufacturing* 77 (2023): 103806. https://doi.org/10.1016/j.addma.2023.103806.

20. Moradi, Mahmoud, Ahmad Aminzadeh, Davood Rahmatabadi, and Alireza Hakimi. "Experimental investigation on mechanical characterization of 3D printed PLA produced by fused deposition modeling (FDM)." *Materials Research Express* 8, no. 3 (2021): 035304.

21. Xiao, Xianglian, Venkata S. Chevali, Pingan Song, Dongning He, and Hao Wang. "Polylactide/hemp hurd biocomposites as sustainable 3D printing feedstock." *Composites Science and Technology* 184 (2019): 107887. https://doi.org/10.1016/j.compscitech.2019.107887.

Empowering Sustainability with Statistical Modelling of FDM-Printed Bamboo-Based Green Composites

6

6.1 INTRODUCTION

Fused filament fabrication, also known as fused deposition modelling (FDM), is one of the widely used 3D printing techniques. In FDM, feedstock materials such as metals, ceramics, and polymers are used in the form of filaments. For small-scale industries and applications, polymers are predominantly adopted due to their lightweight nature, lower capital cost, aesthetic appeal, and wide range of availability [1]. Among polymers, thermoplastics and thermosetting polymers are commonly used; however, thermoplastic polymers are preferred

 DOI: 10.1201/9781003732143-6

for additive manufacturing (AM) because they are recyclable and well-suited for such applications.

Moreover, various polymers such as polypropylene (PP), polyethylene terephthalate glycol (PETG), acrylonitrile butadiene styrene (ABS), and poly-lactic acid (PLA) have been used as feedstock materials for FDM to fabri-cate required components. Among these polymers, PLA is the most widely adopted due to its biodegradable and eco-friendly nature compared to other thermoplastic polymers. Kopar and Yildiz [2] studied the mechanical prop-erties, such as ultimate tensile strength (UTS), bending stress, and impact strength of components made from PLA, ABS, and PETG using the FDM process. This study revealed that PLA exhibited higher tensile and bend-ing strengths (47.95 and 56.80 MPa, respectively) compared to the other two polymers. However, it showed a lower impact strength (3.14 kJ/m²) than ABS and PETG.

Despite its advantages, the mechanical, thermal, and morphological properties of PLA still require improvement. Reinforcing PLA with natu-ral fibres and incorporating additives such as coupling agents and tough-ening agents can significantly enhance these properties in comparison to pure PLA. Pereira et al. [3] developed a PLA filament reinforced with rice husk for FDM applications. The results showed that incorporating 5 wt% of rice husks into PLA exhibited good printability, no nozzle clogging, and defect-free components. However, the mechanical properties were reduced compared to virgin PLA. Mohammadsalih et al. [4] investigated the mechanical properties of PLA and PP filaments reinforced with wood fibres for 3D printing applications. The results indicated that the flexural properties improved compared to pure PLA and PP. Additionally, the tensile properties increased with up to 30 wt% of wood filler but declined when the fibre loading (FL) exceeded this threshold. Beg et al. [5] studied the physical and mechanical properties of PLA filaments reinforced with hemp hurds for AM applications. The outcome revealed that the tensile strength improved by 8% compared to virgin PLA.

Furthermore, the thermal properties of the composite specimens were enhanced compared to neat PLA. Kacem et al. [6] developed PLA bio-composites reinforced with Yucca fibres and evaluated the thermal and mechanical properties of hybrid composite components printed by FDM. The results showed that the tensile (22%), compression (20%), and flexural (12%) strengths improved compared to neat PLA. Moreover, the optimization of FDM process parameters is crucial for enhancing the mechanical proper-ties of polymers and polymer composite components printed by the FDM process.

Samykano et al. [7] studied the influence of FDM process parameters on ABS-printed samples' mechanical properties. The results found that the optimal process parameter settings, i.e., layer height 0.5 mm, infill density (ID) 80%, and raster angle 65°, are identified at which the greater mechanical properties like tensile strength 31.57 MPa, modulus of elasticity 774.20 MPa, fracture strain 0.094 mm/mm, and toughness 2.28 J/m^3 were obtained. Afrose et al. [8] investigated the effect of FDM process parameters, i.e., build orientations, on printed parts by using PLA material. The result showed that the maximum tensile strength (i.e., 38.7 MPa) has been obtained in PLA-X orientation as compared to PLA-Y and PLA-45° orientations. Chalgham et al. [9] studied the influence of FDM process parameters on PLA-printed samples' mechanical properties measured via three-point bending test. The results revealed that the best process parameters were identified; layer height is 0.3 mm, nozzle temperature (NT) at 190°C, printing speed at 90 mm/s, and X-Z part build orientations. Samykano [10] explored the effect of FDM process parameters on PLA-printed samples. The results revealed that the optimal process parameter settings, i.e., layer height 0.3 mm, ID 80%, and raster angle 40°, are identified at which the higher mechanical properties such as UTS 28.45 MPa, modulus of elasticity 828 MPa, fracture strain 0.08 mm/mm, and toughness 1.72 J/m^3 were obtained.

Additionally, Alkabbanie et al. [11] conducted a study to determine optimal process parameter setting of FDM process of PLA composites with reinforcement of short carbon fibres. The findings demonstrated that the mechanical properties of PLA composite improved significantly. Specifically, impact strength increased by 14.49% and flexural strength improved by 28.05% under the optimal process parameter setting. The optimal process parameters were identified as a layer height of 0.1 mm, a feed rate of 7 mm^3/s, and an extrusion temperature of 230°C. Yu et al. [12] investigated the properties of PLA components reinforced with polybutylene adipate terephthalate by FDM process. The investigated results showed an 142.97% increase in crystallinity and enhanced thermal stability in the blended compound as compared to virgin PLA. Additionally, the compound exhibited improved hydrophilicity and a decrease in melt flow index values.

Limited research exists on PLA polymers reinforced with natural fibres and coupling agents for FDM printing. These hybrid composites offer biodegradability, eco-friendliness, and enhanced mechanical properties compared to pure PLA components. This chapter focuses on statistical analysis of FDM process parameters on printing of bamboo-based green composites. Experimentation is done using Taguchi (L_9) orthogonal array considering four FDM parameters, namely NT, layer thickness (LT), ID, and FL while UTS and Young's modulus are the responses. Initially, bamboo stems were processed into short fibres (~300 μm), then blended with PLA and additives using a twin-screw internal melt mixer. Further, composite filaments were produced using a single-screw extruder. Finally, FDM-printed specimens were mechanically characterized, followed by statistical analysis conducted to evaluate the effects of process parameters on mechanical properties and identify the optimal FDM process parameters.

6.2 MATERIALS AND METHODS

6.2.1 Thermo-Polymers and Coupling Agents

Bio-PLA with the code KBB1012 was used as the matrix polymer material. Short bamboo fibres (approximately 300 µm in size) were selected as the reinforcement material at different weight percentages, i.e., 3%, 6%, and 9%, respectively. To ensure proper adhesion between the matrix and the reinforcement, coupling agents such as malic anhydride (2 wt%, density 1.48 g/cm^3) and polyethylene glycol (6 wt%, density 1.125 g/cm^3) were adopted [13].

6.2.2 Processing of Bamboo Fibres

Bamboo stems were extracted from bamboo plants, cut into small stakes, and converted into powder form using a shearing operation. The short bamboo fibres were then sieved to maintain a uniform size of approximately 300 µm. Subsequently, the short bamboo fibres were treated with a sodium hydroxide (NaOH) solution to remove non-cellulose constituents, wax, and surface impurities [14]. The chemically treated bamboo fibres were then washed with acetone and deionized water to remove any residual NaOH solution from their surfaces and were dried in a hot air oven at 60°C for 24 hours [13]. The processing of bamboo fibres with sequence of operations is shown in Figure 6.1.

6.2.3 Preparation of PLA Composite Samples by FDM Process

First, a compound mixture of Bio-PLA reinforced with different weight percentages of short bamboo fibres and additives was prepared using a twin-screw melt mixer setup. Subsequently, PLA-based composite filaments reinforced with short bamboo fibres and additives were produced using a single-screw filament extruder (Model: HAAKE Rheomex CTW5, Thermo Fisher, Germany). Tensile specimens (according to ASTM D638) were fabricated using FDM,

FIGURE 6.1   Sequence of operations for processing of bamboo fibres

TABLE 6.1   FDM process parameters and their levels

S. NO.	PROCESS PARAMETER	SYMBOL	LEVELS		
			LEVEL-1	LEVEL-2	LEVEL-3
1	Layer thickness in mm	A	0.1	0.2	0.3
2	Infill density in %	B	60	80	100
3	Nozzle Temperature in °C	C	195	200	205
4	Filler in Wt%	D	3	6	9

Abbreviation: FDM, fused deposition modelling.

with the PLA composite filament serving as the feedstock material. The samples were printed (as shown in Figure 6.2) based on a Taguchi L_9 orthogonal array design, and the corresponding FDM process parameters and their levels are presented in Table 6.1.

6.2.4 Testing Details

Tensile tests on PLA composite specimens and pure PLA specimens were conducted using an INSTRON 3367 machine, as shown in Figure 6.3, which

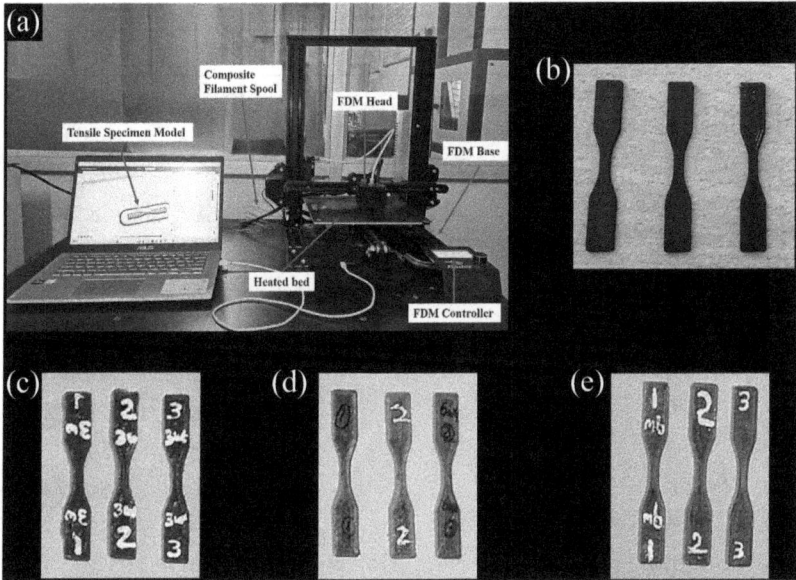

FIGURE 6.2 (a–e) Fabrication of tensile samples by using FDM printer

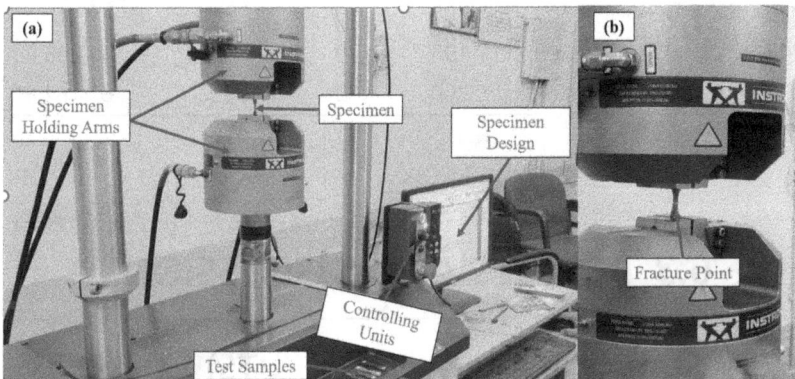

FIGURE 6.3 (a) Universal testing machine setup. (b) Specimen fracture point after testing

has a load capacity of ±20 kN and is equipped with a 40 mm axial extensometer. All tensile tests were performed at a crosshead speed of 1 mm/min, and the corresponding values of UTS, modulus of elasticity (E), and percentage elongation were determined for the samples. Further, three samples were tested for each set of measurements. The tensile test results of PLA composite samples are depicted in Table 6.2.

TABLE 6.2   Tensile test results of PLA composite specimens

N	A	B	C	D	Y1	Y2
1	0.1	60	195	3	15.61	510.32
2	0.1	80	200	6	31.36	698.63
3	0.1	100	205	9	50.05	552.32
4	0.2	60	200	9	50.58	681.44
5	0.2	80	205	3	25.37	732.48
6	0.2	100	195	6	34.88	802.6
7	0.3	60	205	6	35.24	720.08
8	0.3	80	195	9	59.62	513.95
9	0.3	100	200	3	27.85	612.52
10	0.1	60	195	3	14.29	509.42
11	0.1	80	200	6	30.04	697.73
12	0.1	100	205	9	48.73	551.42
13	0.2	60	200	9	49.26	680.54
14	0.2	80	205	3	24.05	731.58
15	0.2	100	195	6	33.56	801.7
16	0.3	60	205	6	33.92	719.18
17	0.3	80	195	9	58.3	513.05
18	0.3	100	200	3	26.53	611.62

6.3 RESULTS AND DISCUSSIONS

6.3.1 Statistical Analysis of y_1 (UTS)

The statistical analysis of y_1 on FDM process parameters is analyzed using regression analysis of variance (ANOVA). The ANOVA results for y_1 are shown in Table 6.3. The results show that the linear (A and D) and interaction effects of A and C are significant at a 5% level of significance (p-value ≤ 0.05) [15]. The other terms/parameters, except parameters highlighted by (*) in Table 6.3 are considered to be statistically insignificant. Additionally, the satisfactory and adequacy and fitness of the model are tested via R^2 and

R^2 (adj). The R^2 value for y_1 is 99.67%, which signifies the experimental data are well-suited and satisfactory with experimental data [16]. The adjusted R^2 value for y_1 is 99.44%, which shows the strong agreement between experimental and projected values. Further, normality of the data is tested through a normal probability plot. The normal probability plot for residual is depicted in Figure 6.4(a). The data points are found closer to the straight line, which indicates the y_1 data is normally distributed [17]. Figure 6.4(b) shows the relationship between residual and predicted fits, and data points signify that residual was structureless.

TABLE 6.3 ANOVA for y_1

SOURCE	DF	ADJ SS	ADJ MS	F-VALUE	P-VALUE
Regression	7	3,160.72	451.531	434.46	0
A	1	53	53.005	51	0*
B	1	0.95	0.948	0.91	0.362
C	1	8.22	8.22	7.91	0.018*
D	1	245.08	245.078	235.81	0*
A × B	1	0.05	0.05	0.05	0.831
A × C	1	53.43	53.427	51.41	0*
B × C	1	0.74	0.743	0.71	0.418
Error	10	10.39	1.039		
Lack-of-fit	1	2.55	2.552	2.93	0.121
Pure error	9	7.84	0.871		
Total	17	3,171.11			

R^2: 99.67%, R^2-adj: 99.44%
Abbreviations: ANOVA, analysis of variance; DF, degrees of freedom; MS, mean square; SS, sum of squares.

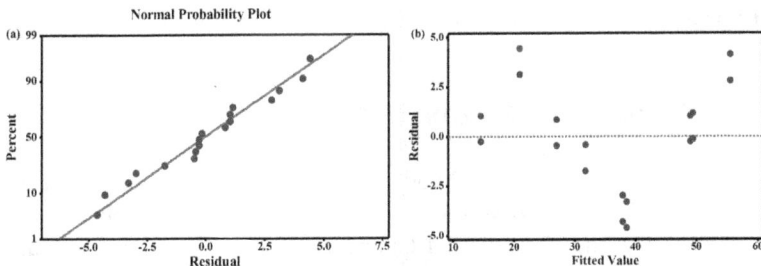

FIGURE 6.4 (a) Normal plot of residuals for y_1. (b) Residual vs. predicted plot for y_1

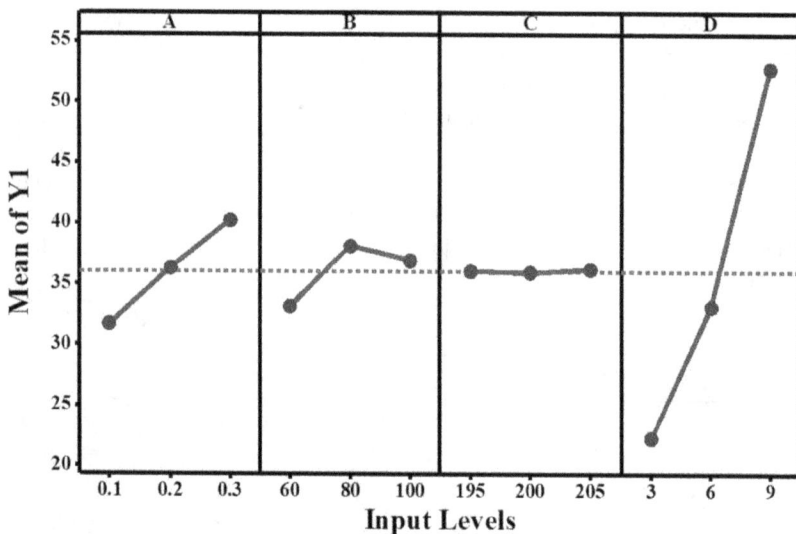

FIGURE 6.5   Main effect plot for y_1

Further, to validate the ANOVA results (Table 6.3), parametric analysis has been performed. The parameters A and D are the significant factors that have the greatest impact on mean of y_1. It was noticed from Figure 6.5, when both parameters A and D vary from 0.1 to 0.3 mm and 3% to 9%, the value of y_1 is drastically increased from 31.68 to 40.24 MPa for factor A and 22.283–52.75 MPa for factor D, respectively. This is due to the fact that higher LT and higher wt% of fibre lead to a stronger composite capable of bearing the external force [18]. On the other hand, the effect of B on y_1 seems to be marginally significant and factor C has no effect on y_1. From the main effect analysis, the optimal setting for the y_1 is obtained as A (0.3 mm, L_3), B (80, L_2), D (9, L_3), and C at any level to get a higher y_1 value.

6.3.2 Statistical Analysis of y_2 (E)

The statistical significance of FDM process parameters on y_2 is analyzed using regression ANOVA. The ANOVA results for y_2 are shown in Table 6.4. The results show that the linear (B and C) and interaction effects of $A \times B$ and $B \times C$ are significant at a 5% level of significance (p-value ≤ 0.05) [13]. The other terms/parameters except parameters highlighted by (*) in Table 6.4 are considered to be statistically insignificant. Additionally, the satisfactory and adequacy and fitness of the model are tested via R^2 and R^2 (adj). The R^2 value

TABLE 6.4   ANOVA for y_2

SOURCE	DF	ADJ SS	ADJ MS	F-VALUE	P-VALUE
Regression	7	167,859	23,980	36.12	0
A	1	643	643	0.97	0.348
B	1	110,794	110,794	166.87	0*
C	1	61,634	61,634	92.83	0*
D	1	58	58	0.09	0.773
A × B	1	7,832	7,832	11.8	0.006*
A × C	1	1,381	1,381	2.08	0.18
B × C	1	106,353	106,353	160.18	0*
Error	10	6.640	664		
Lack-of-fit	1	6,636	6,636	16,385.09	0
Pure error	9	4	0		
Total	17	174,498			

R^2: 96.20%, R^2-adj: 93.53%

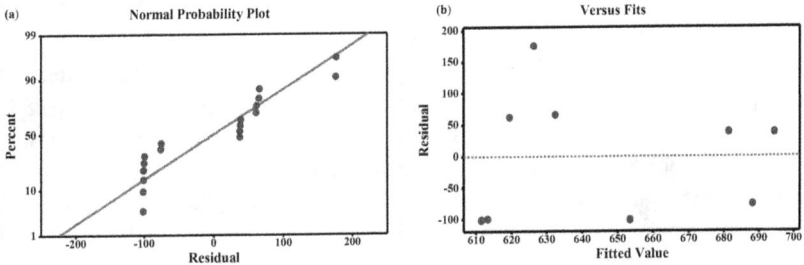

FIGURE 6.6   (a) Normal plot of residuals for y_2. (b) Residual vs. predicted plot for y_2

for y_2 is 96.20%, which signifies the experimental data are well-suited and satisfactory with experimental data [13]. The adjusted R^2 value for y_2 is 93.53%, which shows the strong agreement between experimental and projected values. Further, normality of the data is tested through a normal probability plot. The normal probability plot for residual is depicted in Figure 6.6(a). The data points are found closer to the straight line, which indicates the y_2 data is normally distributed [19]. Figure 6.6(b) shows the relationship between residual and predicted fits, and data points signify that the residual was structureless.

Further, to validate the ANOVA results (Table 6.4), a parametric analysis is done. The parameters B and C are the significant factors that have the greatest impact on mean of y_2 [16]. It was noticed from Figure 6.7, when both

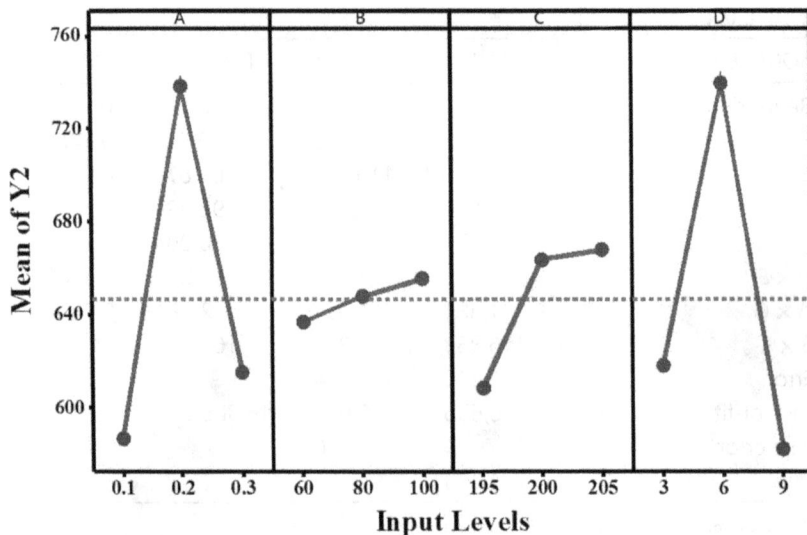

FIGURE 6.7 Main effect plot for y_2

parameters B and C vary from 60% to 100% and 195°C to 205°C, the value of y_2 is increased from 636.83 to 655.363 MPa for factor B and 608.50 to 667.84 MPa for factor C, respectively. This is due to the fact that improved material density and bonding result in a solid structure with enhanced layer adhesion between the matrix and the polymer [20]. On the other hand, the effect of A and D on y_2 seems to be marginally significant because sudden increase and decrease of y_2 is seen in both cases. From the main effect analysis, the optimal setting for y_2 is obtained as A (0.2 mm, L_2), B (80, L_2), C (200, L_2), and D (6, L_2).

6.3.3 Result Validation

The optimal setting obtained from the main effect analysis is done via valida-tion. The optimal settings for y_1: A (0.3 mm, L_3), B (80, L_2), D (9, L_3), and C (195°C, L_1) and for y_2: A (0.2 mm, L_2), B (80, L_2), C (200, L_2), and D (6, L_2) are taken for re-experimentation, and the results of re-experimentation are shown in Table 6.5. The results show that re-experimental values for y_1 and y_2 are found closer to the optimal predicted values. The deviation between the experimental and model results is closer and less than 5%. This signifies that the model results are satisfactory and acceptable [17–20].

TABLE 6.5   Validation results of the FDM process

OPTIMAL SETTING	OUTPUT PARAMETERS	REGRESSION MODEL RESULT	RE-EXP. RESULTS	% DEVIATION
A (0.3 mm), B (80%), C (195°C) and D (9 wt%)	y_1	60.02	59.12	1.4995
A (0.2 mm), B (80%), C (200°C), and D (6 wt%)	y_2	513.45	513.24	0.1517

6.4 CONCLUSION

This chapter presented the statistical analysis of FDM process parameters on printing of bamboo-based green composites. Experimentation is done using Taguchi (L_9) orthogonal array considering four FDM parameters, namely A, B, C, and D while y_1 and y_2 are the responses. The factors A and D (linear effect), $A \times C$ (interaction effect) for y_1 and B and C (linear effect), $B \times C$ (interaction effect) for y_2 are found significant at a 5% level of significance (p-value ≤ 0.05). Similarly, the satisfactory and adequacy and fitness of the model are tested via R_2 and R_2 (adj) and found satisfactory and strong agreement between experimental and predicted values. The best optimal settings obtained are A (0.3 mm), B (80%), D (9 wt%), and C (195°C) for y_1 and A (0.2 mm), B (80%), C (200°C), and D (6 wt%) for y_2. The validation result shows that predicted and re-experimented values for the optimal setting are satisfactory and acceptable. Similarly, no appreciable deviation between predicted and experimental results with a lower average error is seen in the additivity test for the FDM process.

REFERENCES

1. Dubey, Dushyant, Satinder Paul Singh, and Bijoya Kumar Behera. "A review on recent advancements in additive manufacturing techniques." *Proceedings of the Institution of Mechanical Engineers, Part E: Journal of Process Mechanical Engineering* (2024): 1–23. https://doi.org/10.1177/09544089241275860.

2. Kopar, Mehmet, and Ali Riza Yildiz. "Experimental investigation of mechanical properties of PLA, ABS, and PETG 3-d printing materials using fused deposition modelling technique." *Materials Testing* 65, no. 12 (2023): 1–10. https://doi.org/10.1515/mt-2023-0202.

3. Pereira, Daniel F., A. C. Branco, Ricardo Cláudio, Ana C. Marques, and C. G. Figueiredo-Pina. "Development of composites of PLA filled with different amounts of rice husk fibres for fused deposition modelling." *Journal of Natural Fibers* 20 (2023): 1–16. https://doi.org/10.1080/15440478.2022.2162183.

4. Mohammadsalih, Zaid G., Muhammad Muawwidzah, Vasi Uddin Siddiqui, and S. M. Sapuan. "Mechanical properties of wood fibre filled polylactic acid (PLA) composites using additive manufacturing techniques." *Journal of Natural Fiber Polymer Composites* 2, no. 2 (2023): 2821–3289.

5. Beg, Mohammad Dalour Hossen, Kim L. Pickering, John O. Akindoyo, and Christian Gauss. "Recyclable hemp Hurd fibre reinforced PLA composites for 3D printing." *Journal of Materials Research and Technology* 33 (2024): 4439–4447. https://doi.org/10.1016/j.jmrt.2024.10.082.

6. Kacem, Mohamed Amine, Moussa Guebailia, Mohammadreza Lalegani Dezaki, Said Abdi, Nassila Sabba, Ali Zolfagharian, and Mahdi Bodaghi. "Development and 3D printing of PLA bio-composites rein forced with short yucca fibers and enhanced thermal and dynamic mechanical performance." *Journal of Materials Research and Technology* 36 (2025): 1243–1258.

7. Samykano, M., S. K. Selvamani, K. Kadirgama, W. K. Ngui, G. Kanagaraj, and K. Sudhakar. "Mechanical property of FDM printed ABS: Influence of printing parameters." *International Journal of Advanced Manufacturing Technology* 102 (2019): 2779–2796. https://doi.org/10.1007/s00170-019-03313-0.

8. Afrose, Mst Faujiya, S. H. Masood, Pio Iovenitti, Mostafa Nikzad, and Igor Sbarski. "Effects of part build orientations on fatigue behaviour of FDM processed PLA material." *Progress in Additive Manufacturing* 1 (2016): 21–28. https://doi.org/10.1007/s40964-015-0002-3.

9. Chalgham, Ali, Andrea Ehrmann, and Inge Wickenkamp. "Mechanical properties of FDM printed PLA parts before and after thermal treatment." *Polymers (Basel)* 13, no. 8 (2021): 1239. https://doi.org/10.3390/polym13081239.

10. Samykano, M. "Mechanical property and prediction model for FDM 3D printed polylactic acid (PLA)." *Arabian Journal for Science and Engineering* 46 (2021): 7875–7892. https://doi.org/10.1007/s13369-021-05617-4.

11. Alkabbanie, Rasha, Bulent Aktas, Gokhan Demircan, and Serife Yalcin. "Short carbon fiber-reinforced PLA composites: Influence of 3D-printing parameters on the mechanical and structural properties." *Iranian Polymer Journal* (English Ed) 33 (2024): 1065–1074. https://doi.org/10.1007/s13726-024-01315-8.

12. Yu, Wangwang, Mengya Li, Wen Lei, and Yong Chen. "FDM 3D printing and properties of PBAT/PLA blends." *Polymers (Basel)* 16, no. 8 (2024): 1140. https://doi.org/10.3390/polym16081140.

13. Adapa, Sathish Kumar, Jagadish, and Srinivasu Gangi Setti. "Optimization of twin screw melt mixer setup pro cess parameters for better blending of polymers and polymer composites for FDM applications." *Next Materials* 8 (2025): 100671. https://doi.org/10.1016/j.nxmate.2025.100671.

14. Singh, Ashangbam Satyavrata, Sudipta Halder, Jialai Wang, and Jagadish. "Extraction of bamboo micron fibers by optimized mechano-chemical process using a central composite design and their surface modification." *Materials Chemistry and Physics* 199 (2017): 23–33. https://doi.org/10.1016/j.matchemphys.2017.06.040.

15. Mirza, Faizaan, Satish Baloor Shenoy, Srinivas Nunna, Chandrakant Ramanath Kini, and Claudia Creighton. "Effect of material extrusion process parameters on tensile performance of pristine and discontinuous fibre rein forced PLA composites: A review." *Progress in Additive Manufacturing* 10 (2025): 3251–3265. https://doi.org/10.1007/s40964-024-00825-4.

16. Patil, Milind, Mugdha Dongre, D. N. Raut, and Ajinkya Naik. "Multi-objective optimization of fused filament fabrication (FFF) parameters for rice husk reinforced PLA composites." *Next Materials* 8 (2025): 100540. https://doi.org/10.1016/j.nxmate.2025.100540.

17. Muthu Natarajan, S., S. Senthil, and Pandiarajan Narayanasamy. "Investigation of mechanical properties of FDM-processed acacia concinna–filled polylactic acid filament." *International Journal of Polymer Science* 2022, no. 1 (2022): 4761481. https://doi.org/10.1155/2022/4761481.

18. Turkoglu, Turker, and Ahmet Cagri Kilinc. "Optimization of process parameters for steel wire reinforced polylactic acid composites produced by additive manufacturing." *Polymers* 17, no. 5 (2025): 624.

19. Mani, M., A.G. Karthikeyan, K. Kalaiselvan, P. Muthusamy, and P. Muruganandhan. "Optimization of FDM 3-D printer process parameters for surface roughness and mechanical properties using PLA material." *Materials Today: Proceedings* 66 (2022): 1926–1931. https://doi.org/10.1016/j.matpr.2022.05.422.

20. Mohamed, Omar A., Syed H. Masood, and Jahar L. Bhowmik. "Optimization of fused deposition modelling process parameters: A review of current research and future prospects." *Advances in Manufacturing* 3 (2015): 42–53. https://doi.org/ 10.1007/s40436-014-0097-7.

Index